Lecture Notes in Economics and Mathematical Systems

Managing Editors: M. Beckmann and W. Krelle

299

Kurt Marti

Descent Directions and Efficient Solutions in Discretely Distributed Stochastic Programs

Springer-Verlag
Berlin Heidelberg New York London Paris Tokyo

Author

Prof. Dr. Kurt Marti
Universität der Bundeswehr München, Fakultät für Luft- und Raumfahrttechnik
Werner-Heisenberg-Weg 39, D-8014 Neubiberg, FRG

ISBN 3-540-18778-2 Springer-Verlag Berlin Heidelberg New York
ISBN 0-387-18778-2 Springer-Verlag New York Berlin Heidelberg

Printed in Germany

Printing and binding: Druckhaus Beltz, Hemsbach/Bergstr.
2142/3140-543210

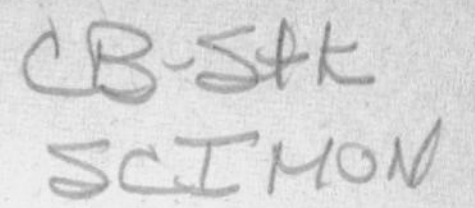

INTRODUCTION

Many problems in stochastic optimization can be represented by

minimize $Eu(A(\omega)x-b(\omega))$ s.t. $x \in D$, (i)

where $(A(\omega),b(\omega))$ is an $m x(n+1)$ random matrix, "E" denotes the expectation operator, and the feasible domain D of (1) is a convex subset of $\mathbb{R}^n$. Moreover, $u:\mathbb{R}^m \to \mathbb{R}$ designates a convex loss function on $\mathbb{R}^m$ measuring the loss arising from the deviation $z=A(\omega)x-b(\omega)$ between the output $A(\omega)x$ of a stochastic linear system $x \to A(\omega)x$ and a random target m-vector $b(\omega)$. Several concrete examples are mentioned in section 1, see also [17],[22],[28],[59],[61],[67].

I. Difficulties in solving problem (i)

Having to solve a mean value minimization problem of the above type, in practice one meets the following considerable difficulties:

I.1. Multiple integrals in (i)

Under weak assumptions [24],[33],[64], the gradient or sub/quasi-gradient of the objective function

$F(x) = Eu(A(\omega)x-b(\omega))$ (ii)

exists and has the form

$\nabla F(x) = EA(\omega)'\nabla u(A(\omega)x-b(\omega))$, (iii)

where ∇u is the gradient or a sub/quasi-gradient of u, and A' denotes the transpose of a matrix A. Corresponding formulas hold also for the higher derivatives of F, cf. [37].

Consequently, in the present case, the derivatives $\nabla F, \nabla^2 F, \ldots$ of the mean value function F are defined, in general, by certain multiple integrals. Thus, any standard mathematical programming routine, based on derivatives of F, is not very useful in solving (i) since multiple integrals can be computed only with a big computational effort.

There are two main types of procedures to overcome this difficulty.

I.1.1. Approximations of the objective function F

Approximations of F can be obtained by

- approximations (e.g. discretizations) of the probability distribution $P_{(A(\cdot),b(\cdot))}$ of the random matrix $(A(\omega),b(\omega))$, see e.g. [4],[25],[26],[30],[37],[63], and by
- approximations of the loss function u, see [5],[29],[37].

If the loss function u is a convex polyhedral function, see section 2, and $(A(\omega),b(\omega))$ has a discrete distribution, then (i) can be represented by a linear program having a dual block angular matrix structure. Hence, the discretization of $P_{(A(\cdot),b(\cdot))}$ is a very attractive method, in this case, for solving (i), see [19], [27],[46],[63]. Unfortunately, the refinement of the discretization of $P_{(A(\cdot),b(\cdot))}$ yields very large scale linear programs. The size of these approximating linear programs can be controlled to some extent by searching for problem-specific discretizations of $P_{(A(\cdot),b(\cdot))}$, or by special refining strategies, cf. [27],[39],[56], [60],[65],[66].

I.1.2. Stochastic approximations of the gradient (sub/quasi-gradient) ∇F

Applying stochastic approximation methods to (iii), we obtain e.g. the stochastic gradient procedure [15],[23],[33],[59]

$$X_{t+1} = p_D(X_t-\rho_t A(\omega_t)'\nabla u(A(\omega_t)X_t-b(\omega_t))),\ t=1,2,\ldots, \qquad \text{(iv)}$$

where $(A(\omega_t),b(\omega_t))$, $t=1,2,\ldots$, is a sequence of independent realizations of $(A(\omega),b(\omega))$, $\rho_t>0$ is a step size, ∇u denotes the gradient or a sub/quasi-gradient of u, and p_D designates the projection operator from $\mathbb{R}^n$ onto D.

Under mild assumptions the sequence $(X_t(\omega))$ defined by algorithm

(iv) converges [62] with probability 1 to the set D^* of optimal solutions of (i). Unfortunately, due to their probabilistic nature, stochastic approximation procedures have only a very slow asymptotic convergence rate of the type

$$E||X_t(\omega)-x^*||^2 = O(t^{-\lambda}) \text{ as } t\to\infty$$

with some constant $0<\lambda\leq 1$. However, the main disadvantage of procedures of the type (iv) is their nonmonotonicity which may be displayed sometimes in a highly oscillatory behavior. Hence, in many cases, one does not know whether the algorithm has already reached a certain neighbourhood of an optimal solution x^* of (i) or not, cf. [14].

Consequently, based mainly on some adaptive step size selections, several methods were suggested [31],[47],[48],[49] for improving the convergence behavior of stochastic approximation procedures. A further method for accelerating algorithms of the type (iv) is based, cf. [41],[45], on the possibility of constructing (feasible) descent directions of F at certain iteration points, see page VII.

I.2. Uncertainty of the proper selection of the loss function u

The second main difficulty is caused by the fact that, in practice, the penalty costs involved in the loss function u, measuring the deviation $z=A(\omega)x-b(\omega)$ between the output $A(\omega)x$ of the stochastic linear system $x \to A(\omega)x$ and the random target $b(\omega)$, cannot often be specified accurately. Hence, in practice, there is always some uncertainty concerning the proper selection of the true loss function u_o. Thus, if u is not properly chosen, then e.g. the direction being calculated according to formula (iii) may be far away from the true gradient given by (iii) for $u=u_o$.

This difficulty is handled in portfolio optimization [36],[61] by introducing the notion of efficient portfolios. On the other hand, depending on the type of the distribution $P_{(A(\cdot),b(\cdot))}$ of $(A(\omega),b(\omega))$, in [38],[40] the special structure of (i) is used to compute feasible descent directions which are stable with respect to variations of the loss function u. These methods are based on stochastic dominance concepts, see e.g. [6],[21],[51],[55].

I.3. Uncertainty of the proper selection of $P_{(A(\cdot),b(\cdot))}$

The third main difficulty results from the fact that, in practice, the probability distribution $P_{(A(\cdot),b(\cdot))}$ of $(A(\omega),b(\omega))$ is not known exactly. Basically, there are two methods to handle this difficulty:

I.3.1. Minimax approach

Assuming that the unknown distribution $P_{(A(\cdot),b(\cdot))}$ is an element of a given set Λ of probability distributions on $\mathbb{R}^{m\cdot(n+1)}$, problem (i) is replaced by

$$\text{minimize } \sup_{\lambda\in\Lambda} \int u(Ax-b)\lambda(dA,db) \quad \text{s.t. } x \in D, \tag{v}$$

cf. [1],[8],[13],[16]. Having a least favorable distribution γ, see e.g. [17], problem (v) is again reduced to a problem of the type (i), cf. [40].

I.3.2. Estimation of $P_{(A(\cdot),b(\cdot))}$

Difficulty (I.3) is handled very often by replacing the unknown distribution $P_{(A(\cdot),b(\cdot))}=\lambda$ in (i) by a certain estimate β of λ, cf. [34],[40],[57]. Mostly, the estimate β of λ is calculated by using a standard statistical estimation procedure taking not into consideration that the outcome β of the estimation process is used in a subsequent decision making problem, see [11],[50].
Estimation methods minimizing the increase of loss, due to the use of an estimation β instead of the true distribution λ of $(A(\omega),b(\omega))$, are considered in [40].

II. Purpose of the work

The purpose of the present work is to give a contribution to the handling of the above mentioned difficulties (I.1)-(I.3). The presented methods work under the assumption that, at least, the random matrix $A(\omega)$ has a discrete distribution. The considerations may be divided into the following two parts.

II.1. Construction of feasible descent directions of F

Using the special structure of (i), we are going to construct feasible descent directions h of F at certain points $x \in \mathbb{R}^n (x \in D)$ such that

- * h can be computed without using any derivatives of F
- * h is stable with respect to variations of the loss function u in a large class U of convex functions on $\mathbb{R}^m$.

Note that safeguarding against variations of u is closely related to the stochastic dominance considerations in many other areas of decision theory and statistics.

The feasible descent directions h can be used in practice as follows:

II.1.1. Semi-stochastic approximation procedures

Stochastic approximation procedures, e.g. the stochastic gradient procedure (iv), can be accelerated very much by using a feasible descent direction h_t given at certain iteration points $X_t, t \in N$, instead of the negative stochastic gradient $-A(\omega_t)'\nabla u(A(\omega_t)X_t - b(\omega_t))$ of F at X_t which is a descent direction only in the mean. Estimates of the increase of the convergence rate are given in [45].

II.1.2. Hybrid algorithms

More generally, the descent directions h computed in the following may be used in any given algorithm for solving (i) for

- omitting the time-consuming computation of derivatives of F at certain iteration points

- improving the convergence behavior of the given algorithm
- stabilizing the algorithm with respect to variations of the loss function u.

II.1.3. Improvements of approximative optimal solutions of (i)

In practice, engineers often have a certain approximation $\tilde{x}$ of an optimal solution x^* of (i). Having at $\tilde{x}$ a feasible descent direction h of F, then, an improvement $\bar{x}$ of $\tilde{x}$, i.e. a point $\bar{x} \in D$ such that $F(\bar{x})<F(\tilde{x})$, may be computed.

Closely related to the construction of descent directions h for F is the computation of stationary (efficient) points:

II.2. Computation of stationary (efficient) points of (i)

If the construction of a feasible descent direction h of F fails at a point $x \in D$, then x is called a D-stationary point of (i). The set S_D of all D-stationary points of (i) - which is computed without using any derivatives of F - includes the optimal solutions x^* of (i) for a large class U of convex loss functions u. A point $x \in S_D$ is efficient in the sense that for x there is no point $\tilde{x} \in D$ such that the distribution $P_{A(\cdot)x-b(\cdot)}$ of $A(\omega)x-b(\omega)$ is stochastically dominated by $P_{A(\cdot)\tilde{x}-b(\cdot)}$, i.e. there is no $\tilde{x} \in D$ such that

$F(\tilde{x}) \leq F(x)$ for all $u \in U$ and

$F(\tilde{x}) < F(x)$ for at least one $\tilde{u} \in U$,

where $F=F_u$ is the objective function of (i), cf. (ii).

Thus, not knowing the proper loss function u_o, the decision maker may use the elements x of S_D as substitutes for an optimal solution $x^*(u_o)$ of (i) which can not be computed then.

On the other hand, even if the loss function $u \in U$ is exactly known, the set S_D of stationary points of (i) has the following practical meaning in solving (i):

II.2.1. Necessary optimality condition

Each candidate $\tilde{x}$ for an optimal solution of (i) must fulfill the constraint

$$\tilde{x} \in S_D, \qquad \text{(vi)}$$

hence (vi) yields derivative-free informations about the location of optimal solutions of (i).

II.2.2. Sharper restrictions for (i)

Knowing S_D, the constraint "$x \in D$" in (i) may be replaced by the sharper constraint (vi), hence, (i) can be replaced by

$$\min Eu(A(\omega)x-b(\omega)) \quad \text{s.t.} \quad x \in S_D. \qquad \text{(i)'}$$

ACKNOWLEDGMENTS

I thank Frau E. Herchenröder for the exellent typing of the manuscript. I thank Frau S. Haussmann for her linguistic support. Finally, I thank Springer-Verlag for the advice in preparing the manuscript and for including the work in the Springer Lecture Notes Series.

München
November 1987

Kurt Marti

CONTENTS

Section | Page

Section Page

Section Page

1. Stochastic programs with a discrete distribution

In stochastic optimization, see e.g. [20],[33],[53], we often have to minimize a convex mean value function of the type

$$F(x) = Eu(A(\omega)x-b(\omega)),\ x \in \mathbb{R}^n \tag{1}$$

subject to the constraint $x \in D$. Here, D is a convex subset of $\mathbb{R}^n$, $u: \mathbb{R}^n \to \mathbb{R}$ is a convex loss function on $\mathbb{R}^m$, $(A(\omega),b(\omega))$ is a random $(m,n+1)$-matrix, and "E" denotes the expectation operator. Several practical problems may be formulated within this framework.

Example. For stochastic linear programs with recourse we have that

$$(A(\omega),b(\omega)) = \begin{pmatrix} c_o(\omega)' & 0 \\ A_o(\omega) & b_o(\omega) \end{pmatrix}$$

and $u\begin{pmatrix} t \\ \tilde{z} \end{pmatrix} = t+p(\tilde{z})$, where $(A_o(\omega),b_o(\omega),c_o(\omega))$ are the random data of a stochastic linear program, $t = c_o(\omega)'x$ and p is a sublinear function, see e.g. [27], measuring the costs $p(\tilde{z})$ arising from the deviation $\tilde{z} = A_o(\omega)x-b(\omega)$ between $A_o(\omega)x$ and $b_o(\omega)$.

Further examples are: Optimal portfolio selection [39],[51], error minimization and optimal design problems [45],[50], statistical prediction [2].

In this book we work under the following two different distribution assumptions (PD1) and (PD2):

In the first case we suppose that the random matrix $(A(\omega),b(\omega))$ has a joint discrete probability distribution

$$P_{(A(\cdot),b(\cdot))} = \sum_{i \in R} \alpha_i \, \varepsilon_{(A^i,b^i)} \tag{PD1}$$

with a finite or countably infinite spectrum (A^i,b^i), $i \in R$, i.e. $R = \{1,2,\dots,r\}$, $r \in \mathbb{N}$, or $R = \mathbb{N} = \{1,2,\dots\}$, $\alpha_i>0$ for all $i \in R$, $\sum_{i\in R} \alpha_i=1$, and $\varepsilon_{(A^i,b^i)}$ denotes the one-point measure in the given $(m,n+1)$-matrix (A^i,b^i). Hence, here it is

$$F(x) = \sum_{i\in R} \alpha_i u(A^i x-b^i) \quad x \in \mathbb{R}^n.$$

In the second case we assume that $A(\omega)$ has a discrete distribution

$$P_{A(\cdot)} = \sum_{i\in R} \alpha_i \varepsilon_{A^i} \tag{PD2}$$

with a finite or countably infinite spectrum A^i, $i \in R$.

According to [27], the objective function F is then given by

$$F(x) = \sum_{i\in R} \alpha_i \bar{u}_i(A_i x), \quad x \in \mathbb{R}^n,$$

where for each $i \in R$ the convex function $\bar{u}_i=\bar{u}_i(z)$ is defined by the conditional expectation

$$\bar{u}(z|A^i) = E(u(z-b(\omega))|A(\omega) = A^i), \quad z \in \mathbb{R}^m.$$

We suppose that all expectations arising in the following are defined (at least) in the L_1-sense.

The problem is now, for given $x \in \mathbb{R}^n$, $x \in D$, resp., to find a vector $y \in \mathbb{R}^n$, $y \in D$, resp., such that $F(y)\leq(<)F(x)$. We mention the following obvious, but important property of the convex optimization problem

$$\text{minimize } F(x) \text{ subject to } x \in D. \tag{SOP}$$

Lemma 1.1. a) If $F(y)<F(x)$, then $F(x+\lambda(y-x))<F(x)$ for all $0<\lambda<1$. If $F(y)=F(x)$ and F is not constant on the line segment xy joining x and $y\neq x$, then $F(x+\lambda(y-x))<F(x)$ for all $0<\lambda<1$. Hence, in both cases $h=y-x$ is a direction of decrease of F in x; if in addition x, $y \in D$, then h is a feasible descent direction of F at x with respect to D, i.e. there exists a positive number t_0 such that $x+th \in D$ and $F(x+th)<F(x)$ for all $0<t<t_0$.

2. Stochastic dominance (SD) and the construction of feasible descent directions

For two probability measures μ,ν on $\mathbb{R}^m$ and a set U of convex functions u on $\mathbb{R}^m$ the stochastic dominance (SD) relations $\nu \precsim_U (\prec_U) \mu$ between μ,ν are defined by

$\nu \precsim_U (\prec_U) \mu$ if and only if $\int u d\nu \leq (<) \int u d\mu$ for all $u \in U$

[6,16,47]. If P_x denotes the probability distribution of $A(\omega)x-b(\omega)$, $A(\omega)x$, resp., hence $F(x)=\int u dP_x$, then $P_y \precsim_U (\prec_U) P_x$ implies that $P_{x+\lambda(y-x)} \precsim_U P_x$ or $P_{x+\lambda(y-x)} \prec_U P_x$ for all $0<\lambda<1$.

Now, let

$\mu = \sum_{i \in R} \mu_i \varepsilon_{z^i}$ and $\nu = \sum_{j \in T} \nu_j \varepsilon_{w^j}$ be two arbitrary probability distributions on $\mathbb{R}^m$ having finite or-countably infinite spectrums $z^i, i \in R$, $w^j, j \in T$, resp., and $\mu_i>0$, $\nu_j>0$ for all i,j. We consider now a finite or infinite matrix $\Pi = (\pi_{ij})$ having elements π_{ij}, $i \in R, j \in T$, which satisfy the following (SD-) conditions

(SD1) $\quad \sum_{j \in T} \pi_{ij} = 1, \ \pi_{ij} \geq 0$ for all $i \in R$, $j \in T$,

(SD2) $\quad \nu_j = \sum_{i \in R} \pi_{ij} \mu_i$ for all $j \in T$,

(SD3) $\quad w^j = \sum_{i \in R} \frac{\mu_i \pi_{ij}}{\nu_j} z^i$ for all $j \in T$.

According to (SD1) Π is a stochastic matrix, and from (SD1) and (SD3) we have that

$$\int z \nu(dz) = \sum_{j \in T} \nu_j w^j = \sum_{i \in R} \mu_i z^i = \int z \mu(dz).$$

By Jensen's inequality and Hardy-Littlewood-Polya-Blackwell-Sherman [9],[19],[43] we obtain the following result concerning the means $\int f d\mu = \sum_{i \in R} \mu_i f(z^i)$, $\int f d\nu = \sum_{j \in T} \nu_j f(w^j)$

of convex functions f on $\mathbb{R}^m$.

Theorem 2.1. a) The relations (SD1)-(SD3) imply that

$$\int f d\nu \leq \int f d\mu \tag{2}$$

for all convex functions f on $\mathbb{R}^m$ which are μ - as well as ν - integrable. b) Let μ,ν have finite spectrums. Then (2) holds for all convex function f on $\mathbb{R}^m$ if and only if the relations (SD1)-(SD3) are fulfilled.

Since in (2) we should also have the strict inequality "<", in addition to (SD1)-(SD3) we still need this next relation between μ and ν:

(SD4a) For at least one $j \in T$ the transition probability measure

$$K^j = \sum_{i \in R} \frac{\mu_i \pi_{ij}}{\nu_j} \varepsilon_{z^i}$$

is not a one-point measure, i.e. K^j has not concentrated its unit-mass in a single point.

In some cases (SD4a) may be replaced by the stronger condition.

(SD4b) There exists at least one $j \in T$ such that K^j is not a one-point measure and $\pi_{ij}>0$ for all $i \in R$.

Now we have this result.

Theorem 2.1.1. a) If μ,ν satisfy (SD1)-(SD4a), then

$$\int f d\nu < \int f d\mu \tag{2'}$$

for all convex functions f on $\mathbb{R}^m$ which are μ - as well as ν - integrable and strictly convex on the convex hull $\text{conv}\{z^i : i \in R\}$ of the m-vectors $z^i, i \in R$. b) If (SD1)-(SD3) and (SD4b) are fulfilled, then (2') holds for all convex functions f on $\mathbb{R}^m$ which are μ - as well as ν - integrable and which are not affine-linear on $\text{conv}\{z^i : i \in R\}$.

Proof. The strict inequality (2') follows from

$$\int f d\nu = \sum_{j\in T} \nu_j f(w^j) = \sum_{j\in T} \nu_j f\Big(\sum_{i\in R} \frac{\mu_i \pi_{ij}}{\nu_j} z^i\Big)$$

$$< \sum_{j\in T} \nu_j \sum_{i\in R} \frac{\mu_i \pi_{ij}}{\nu_j} f(z^i) = \sum_{i\in R} \mu_i f(z^i) = \int f d\mu,$$

where in the proof of the second part we use, in addition, the fact that the equation $f(\sum_{i\in R} \beta_i z^i) = \sum_{i\in R} \beta_i f(z^i)$, with $\beta_i > 0$ for all $i \in R$ and $\sum_{i\in R} \beta_i = 1$, implies that $f(\sum_{i\in R} \lambda_i z^i) = \sum_{i\in R} \lambda_i f(z^i)$ for all sequences (λ_i) such that $\lambda_i \geq 0, i \in R$ and $\sum_{i\in R} \lambda_i = 1$.

Obviously, the above considerations may be applied to our problem described in section 1.

2.1. SD-conditions (3.1)-(3.3),(3.4a),(3.4b),(3.5) for the construction of feasible descent directions (distribution assumption (PD1)). Suppose that $P_{(A(\cdot),b(\cdot))} = \sum_{i\in R} \alpha_i \varepsilon_{(A^i,b^i)}$, and let x be any given n-vector; furthermore, let y=y(x) denote a yet unknown vector such that $F(y) \leq (<) F(x)$. Define

$z^i = A^i x - b^i$, $i \in R$,

$w^j = A^j y - b^j$, $j \in T = R$;

clearly, some of the vectors $z^i, i \in R$ resp. $w^j, j \in R$, may be equal. We find that

$$P_{A(\cdot)x-b(\cdot)} = \sum_{i\in R} \alpha_i \varepsilon_{(A^i x - b^i)} = \sum_{i\in R} \alpha_i \varepsilon_{z^i},$$

$$P_{A(\cdot)y-b(\cdot)} = \sum_{j\in R} \alpha_j \varepsilon_{(A^j y - b^j)} = \sum_{j\in R} \alpha_j \varepsilon_{w^j}.$$

Consequently, for $\mu = P_{A(\cdot)x-b(\cdot)}$ and $\nu = P_{A(\cdot)y-b(\cdot)}$ conditions (SD1)-(SD4a),(SD4b) can be written in the following form:

(3.1) $\sum_{j \in R} \pi_{ij} = 1,\ \pi_{ij} \geq 0$ for all $i,j \in R$,

(3.2) $\alpha_j = \sum_{i \in R} \alpha_i \pi_{ij}$ for all $j \in R$,

(3.3) $A^j y - b^j = \sum_{i \in R} \frac{\alpha_i \pi_{ij}}{\alpha_j}(A^i x - b^i)$ for all $j \in R$,

(3.4a) The transition probability measure

$$K^j = \sum_{i \in R} \beta_{ij} \varepsilon_{z^i},\ \beta_{ij} := \frac{\alpha_i \pi_{ij}}{\alpha_j},$$

is not a one-point measure for at least one $j \in R$.

(3.4b) There exists at least one $j \in R$ such that K^j is not a one-point measure and $\pi_{ij} > 0$ for all $i \in R$.

We observe that for given $x \in \mathbb{R}^n$ (3.1)-(3.3) is a system of linear equalities/inequalities for the determination of the unknowns y, Π, where $y \in \mathbb{R}^n$ and $\Pi = (\pi_{ij})_{i,j \in R}$ is a finite or infinite stochastic matrix. Note that (3.1)-(3.3) has always the trivial solution $y=x$, $\Pi=I$, where I is the identity matrix of size $|R|$.

Let $\bar{A} = EA(\omega) = \sum_{i \in R} \alpha_i A^i$, $\bar{b} = Eb(\omega) = \sum_{i \in R} \alpha_i b^i$ denote the mean values of $A(\omega), b(\omega)$ respectively. If (y,Π) is a solution of (3.1)-(3.3), then

$$\bar{A}y - \bar{b} = E(A(\omega)x - b(\omega)) = \sum_{j \in R} \alpha_j (A^j y - b^j)$$

$$= \sum_{j \in R} \alpha_j \Big(\sum_{i \in R} \frac{\alpha_i \pi_{ij}}{\alpha_j}(A^i x - b^i)\Big) = \sum_{i \in R} \alpha_i (A^i x - b^i)$$

$$= E(A(\omega)x - b(\omega)) = \bar{A}x - \bar{b},$$

hence we proved this lemma:

Lemma 2.1. If for given $x \in \mathbb{R}^n$ the tuple (y,Π) is a solution of (3.1)-(3.3), then y,x must satisfy the linear relation

$\bar{A}y = \bar{A}x.$ (4)

Clearly, (4) has for each $x \in \mathbb{R}^n$ at least one solution y.

In order to take also into consideration the constraint "$x \in D$" of the original stochastic optimization problem (SOP), where D is a convex subset of $\mathbb{R}^n$, to the above conditions (3.1)-(3.4) we may still add this condition

(3.5) $y \in D$.

Denote by U_P all convex functions $u: \mathbb{R}^m \to \mathbb{R}$ such that the random variable $u(A(\omega)x-b(\omega))$ is P-integrable for each $x \in \mathbb{R}^n$; if $P_{(A(\cdot),b(\cdot))}$ has a bounded support, then U_P contains all convex functions u on $\mathbb{R}^m$. From Theorem 2.1 and 2.1.1 we now obtain the next result, a preliminary version of it was already given in [31],[35].

Theorem 2.2. Let x be a given n-vector. a) If (y,Π) is a solution of (3.1)-(3.3), then $F(y) \leq F(x)$ for each convex function $u \in U_P$. b) If (y,Π) solves (3.1)-(3.4a), then $F(y)<F(x)$ for every loss function $u \in U_P$ being strictly convex on $\mathrm{conv}\{z^i : i \in R\}$. c) If (3.1)-(3.4b) are fulfilled, then $F(y)<F(x)$ for every $u \in U_P$, which is not affine-linear on $\mathrm{conv}\{z^i : i \in R\}$.

For our (SOP) Lemma 1.1 yields then: If $F(y)<F(x)$, then $h=y-x$ is a descent direction of F at x; if $F(y)=F(x)$, $y \neq x$, then $h=y-x$ is also a descent direction of F at x, provided that F is not constant on $\overline{xy}$. If, in addition, $x \in D$, and if we still obey the condition (3.5), then $h=y-x$ is a feasible descent direction.

2.2. SD-conditions (5.1)-(5.3),(5.4a),(5.4b),(5.5) for the construction of feasible descent directions (distribution assumption (PD2)). Suppose that $P_{A(\cdot)} = \sum_{i \in R} \alpha_i \varepsilon_{A^i}$ and define $\bar{u}(z|A^i) = E(u(z-b(\omega))|A(\omega) = A^i)$, $z \in \mathbb{R}^m$. Hence, the objective function F of (SOP) is given by

$$F(x) = \sum_{i\in R} \alpha_i \bar{u}(A^i x | A^i).$$

Defining here z^i, w^j by

$$z^i = A^i x,\ i \in R,\ w^j = A^j y,\ j \in T = R,$$

we find that

$$P_{A(\cdot)x} = \sum_{i\in R} \alpha_i \varepsilon_{A^i x},\ P_{A(\cdot)y} = \sum_{j\in R} \alpha_j \varepsilon_{A^j y}.$$

Given $x \in \mathbb{R}^n$, corresponding to (SD1)-(SD4) we have to consider here the conditions

(5.1) $\sum_{j\in R} \pi_{ij} = 1,\ \pi_{ij} \geq 0$ for all $i,j \in R$,

(5.2) $\alpha_j = \sum_{i\in R} \alpha_i \pi_{ij}$ for all $j \in R$,

(5.3) $A^j y = \sum_{i\in R} \frac{\alpha_i \pi_{ij}}{\alpha_j} A^i x$ for all $j \in R$,

(5.4a) The transition probability measure

$$K^j = \sum_{i\in R} \beta_{ij} \varepsilon_{z^i},\ \beta_{ij} = \frac{\alpha_i \pi_{ij}}{\alpha_j},$$

is not a one-point measure for at least one $j \in R$,

(5.4b) There exists at least one $j \in R$ such that K^j is not a one-point measure and $\pi_{ij} > 0$ for all $i \in R$.

If $x \in D$, then we may still add the condition

(5.5) $y \in D$.

We observe that (5.1)-(5.4) follow from (3.1)-(3.4) by simply setting $b^i = b^{i_0}$ for all $i \in R$. From (5.1)-(5.3) follows again (4), i.e. $\bar{A}y = \bar{A}x$, moreover, since $\bar{u}(\cdot | A^i)$ is for each $u \in U_P$ and $i \in R$ a convex function on $\mathbb{R}^m$,

$$F(y) = \sum_{j\in R} \alpha_j \bar{u}(A^j y|A^j) = \sum_{j\in R} \alpha_j \bar{u}\Big(\sum_{i\in R} \frac{\alpha_i \pi_{ij}}{\alpha_j} A^i x|A^j\Big)$$

$$\leq \sum_{j\in R} \alpha_j \sum_{i\in R} \frac{\alpha_i \pi_{ij}}{\alpha_j} \bar{u}(A^i x|A^j) \leq \sum_{i\in R} \alpha_i \bar{u}(A^i x|A^i) = F(x),$$

provided that this additional linear inequality for Π holds:

$$(5.6) \quad \sum_{i\in R} \alpha_i \sum_{j\in R} \pi_{ij} \bar{u}(A^i x|A^j) \leq \sum_{i\in R} \alpha_i \bar{u}(A^i x|A^i).$$

If $A(\omega),b(\omega)$ are independent random variables, then (5.6) is satisfied automatically with "=" and $\bar{u}(z|A^i) = Eu(z-b(\omega))$.

Corresponding to Theorem 2.2, in case (PD2) we have now this result:

Theorem 2.3. Let x be a given n-vector. a) If (y,Π) is a solution of (5.1)-(5.3) and (5.6), then $F(y)\leq F(x)$ for each convex function $u \in U_p$. b) If (y,Π) solves (5.1)-(5.4a) and (5.6), then $F(y)<F(x)$ for every loss function $u \in U_p$ such that $\bar{u}(\cdot|A^j)$ is strictly convex on $\mathrm{conv}\{z^i: i \in R\}$ for at least one integer "j" mentioned in (5.4a). c) If (5.1)-(5.4b) and (5.6) hold, then $F(y)<F(x)$ for every $u \in U_p$ such that $\bar{u}(\cdot|A^j)$ is not affine-linear on $\mathrm{conv}\{z^i: i \in R\}$ for at least one integer "j" mentioned in (5.4b).

Note that from the above proposition we obtain of course the same consequences concerning the computation of descent directions in (SOP) as from the related Theorem 2.2.

After this first treatment of the two distribution cases (PD1) and (PD2) considered in this book, we ask now whether, under certain additional assumptions regarding the convex loss function $u \in U_p$, the conditions (3.1)-(3.4) resp. (5.1)-(5.4) may be weakened to some extent. Since (5.1)-(5.4) is included in (3.1)-(3.4) for $b^i=b^{i_o}, i \in R$, it is sufficient to study in the following relations (3.1)-(3.3),(3.4a),(3.4b) only.

2.3. Partly monotone loss. Let $z_I=(z_k)_{k\in K}$, $z_{II}=(z_k)_{k\in K}$ be a partition of $z=(z_1,z_2,\ldots,z_m)'$ and suppose that for any two vectors $z,w\in\mathbb{R}^m$

$$z_I \leq w_I \text{ and } z_{II} = w_{II} \Longrightarrow u(z) \leq(\geq) u(w), \tag{6}$$

where $z_I\leq w_I$ means that $z_k\leq w_k$ for each $k\in K$. In applications, one also finds the following variant of the partial monotonicity property (6):

$$z_I\leq w_I,\ z_{II}=w_{II} \text{ and } z_k<w_k \text{ for } k\in K_1 \Longrightarrow u(z)<(>)u(w), \tag{6a}$$

where K_1 is a subset of K.

Lemma 2.2. If $u\in U_p$ has the partial monotonicity property (6), then Theorem 2.2 remains true if (3.3) and (3.4) are replaced by the weaker linear conditions

$$(3.3)_M \quad \begin{aligned} &(A^j y-b^j)_I \leq (\geq) \sum_{j\in R} \frac{\alpha_i \pi_{ij}}{\alpha_j} (A^i x-b^i)_I,\ j\in R \\ &(A^j y-b^j)_{II} = \sum_{j\in R} \frac{\alpha_i \pi_{ij}}{\alpha_j} (A^i x-b^i)_{II},\ j\in R; \end{aligned}$$

$(3.4a)_M$ There is at least one $j\in R$ such that $K^j_{II} = \sum_{i\in R} \beta_{ij} z^i_{II}$, $\beta_{ij} = \frac{\alpha_i \pi_{ij}}{\alpha_j}$, is not a one-point measure

$(3.4b)_M$ There is at least one $j\in R$ such that K^j_{II} is not a one-point measure and $\pi_{ij}>0$ for all $i\in R$.

Compared to (4), here we have then the weaker condition

$$(\bar{A}y)_I \leq (\geq) (\bar{A}x)_I,\quad (\bar{A}y)_{II} = (\bar{A}x)_{II}. \tag{4a}$$

Obvious extensions of the above lemma are obtained if u has the strict partial monotonicity property (6a). An application of this partial monotonicity case (6a) to stochastic linear programming with recourse is given later on.

2.4. Constant loss on subspaces. Let the loss function u be defined by

$$u(z) = v(Hz), \quad z \in \mathbb{R}^m \tag{7}$$

where H is a fixed $\bar{m}$xm matrix and $v: \mathbb{R}^{\bar{m}} \to \mathbb{R}$ is a convex function. Then

$$F(x) = \sum_{i \in R} \alpha_i \, u(A^i x - b^i) = \sum_{i \in R} \alpha_i \, v(HA^i x - Hb^i).$$

<u>Lemma 2.3.</u> If u is defined by (7), then Theorem 2.2 remains true if u, $(A(\omega), b(\omega))$ is replaced by v, $(HA(\omega), Hb(\omega))$ respectively.

We observe that (3.3) has then the form

$$(3.3)_H \quad H((A^j y - b^j) - \sum_{i \in R} \frac{\alpha_i \, \pi_{ij}}{\alpha_j} (A^i x - b^i)) = 0, \; j \in R$$

and (4) is exchanged for

$$H(\bar{A}y - \bar{A}x) = 0, \tag{4b}$$

which are weaker conditions than the original ones, provided that rank $H<m$.

2.5. <u>Separable loss.</u> Assume that

$$u(z) = u(z_1, \ldots, z_m) = \sum_{k=1}^{m} u_k(z_k), \tag{8}$$

where $u_k: \mathbb{R} \to \mathbb{R}$, $k=1,\ldots,m$, are convex functions on $\mathbb{R}$.
Then it is $F(x) = \sum_{k=1}^{m} F_k(x)$,

$$F_k(x) = Eu_k(A_k(\omega)x - b_k(\omega)) = \sum_{i \in R} \alpha_i \, u_k(A_k^i x - b_k^i),$$

where (A_k, b_k) denotes the k-th row of (A,b). Since the coordinates $A_k^i x - b_k^i$ of $A^i x - b^i$ are separated here by the function u, we may also set up separated SD-relations between the distributions of each pair $A_k(\omega)y - b_k(\omega)$, $A_k(\omega)x - b_k(\omega)$, $k=1,\ldots,m$.
Hence, given $x \in \mathbb{R}^n$, for variables $y \in \mathbb{R}^n$ and $\Pi^{(k)} = (\pi_{ij}^{(k)})_{i,j \in R}$, $k=1,2,\ldots,m$, these relations have the following form:

$$(9.1)_k \quad \sum_{j \in R} \pi_{ij}^{(k)} = 1, \; \pi_{ij}^{(k)} \geq 0, \; i,j \in R,$$

$$(9.2)_k \quad \alpha_j = \sum_{i \in R} \alpha_i \, \pi_{ij}^{(k)}, \; j \in R,$$

$(9.3)_k \quad A_k^j y - b_k^j = \sum_{i\in R} \frac{\alpha_i \pi_{ij}^{(k)}}{\alpha_j} (A_k^i x - b_k^i), \ j \in R,$

$(9.4a)_k \quad K^{j,k} = \sum_{i\in R} \beta_{ij}^{(k)} \varepsilon_{(A_k^i x - b_k^i)}, \ \beta_{ij}^{(k)} = \frac{\alpha_i \pi_{ij}^{(k)}}{\alpha_j}$, is not a one-point measure for at least one $j \in R$,

$(9.4b)_k$ There exists at least one $j \in R$ such that $K^{j,k}$ is not a one-point measure and $\pi_{ij}^{(k)} > 0$ for all $i \in R$.

From these m sets of conditions we get again (4), i.e. $\bar{A}y = \bar{A}x$. Furthermore, if (y,Π) solves (3.1)-(3.4), then $(y,\Pi^{(1)},\ldots,\Pi^{(m)})$ with $\Pi^{(k)}=\Pi$ is a solution of $(9.1)_k$-$(9.4)_k$, $k=1,\ldots,m$.

Corresponding to Theorem 2.2 and 2.3 here we have this result.

Theorem 2.4. Let u be an element of U_P having the form (8) and let x be a given n-vector. a) If $(y,\Pi^{(k)})$ is a solution of $(9.1)_k$-$(9.3)_k$, $k=1,\ldots,m$, then $F(y)\le F(x)$. b) If, in addition to (a), there is at least one integer $k=1,2,\ldots,m$ such that $(9.4a)_k$ holds and u_k is strictly convex on $\text{conv}\{z_k^i: i \in R\}$, then $F(y)<F(x)$. c) If in addition to (a) there is at least one integer $k=1,\ldots,m$ such that $(9.4b)_k$ holds and u_k is not affine-linear on $\text{conv}\{z_k^i: i \in R\}$, then $F(y)<F(x)$.

The same result may be obtained if the convex loss function u is given by $u(z_1,\ldots,z_m) = \prod_{k=1}^{m} u_k(z_k)$ and the rows $(A_k(\omega),b_k(\omega))$, are independent random variables. From the above theorem we obtain again the same consequences concerning the computation of descent directions in (SOP) as from the related Theorem 2.2 and 2.3.

Further special cases allowing weaker SD-conditions may be obtained by combinations of the special cases described in the above sections 2.3-2.5. An important application of section 2.3 is now given as follows.

2.6. Stochastic linear programs with recourse. According to section 1, in this case the loss function u is defined by $u\binom{t}{\tilde z} = t + p(\tilde z)$, $t \in \mathbb{R}$, $\tilde z \in \mathbb{R}^{m-1}$, where p is a sublinear function [27] on $\mathbb{R}^{m-1}$. Furthermore, $(A(\omega),b(\omega))$ has the form

$$(A(\omega),b(\omega)) = \begin{pmatrix} c_o(\omega)' & 0 \\ A_o(\omega) & b_o(\omega) \end{pmatrix}.$$

The partial monotonicity properties (6), (6a) hold then with $K=K_1=\{1\}$. Since $F(x) = Eu(A(\omega)x-b(\omega)) = \bar c_o'x+Ep(A_o(\omega)x-b_o(\omega))$ and $\bar c_o = E\, c_o(\omega)$ is known, it is clear that in the distribution $P_{(A(\cdot),b(\cdot))} = \sum_{i\in R} \alpha_i\, \varepsilon_{(A^i,b^i)}$ we may take $(A^i,b^i) = \begin{pmatrix} \bar c_o & 0 \\ A^i & b^i \end{pmatrix}$, $i \in R$.

Therefore, (3.1),(3.2),$(3.3)_M$,$(3.4)_M$ have then the form

(10.1) $\sum_{j\in R} \pi_{ij} = 1$, $\pi_{ij} \geq 0$, $i,j \in R$,

(10.2) $\alpha_j = \sum_{i\in R} \alpha_i\, \pi_{ij}$, $j \in R$,

(10.3a) $\bar c_o'y \leq (<) \bar c_o'x$,

(10.3b) $A_o^j y - b_o^j = \sum_{i\in R} \frac{\alpha_i\, \pi_{ij}}{\alpha_j} (A_o^i x - b_o^i)$, $j \in R$,

(10.4a) There is at least one $j \in R$ such that $K_o^j = \sum_{i\in R} \beta_{ji}\, \varepsilon_{(A_o^i x - b_o^i)}$

$\beta_{ij} = \frac{\alpha_i\, \pi_{ij}}{\alpha_j}$, is not a one-point measure,

(10.4b) There is at least one $j \in R$ such that K_o^j is not a one-point measure and $\pi_{ij} \geq 0$ for all $i \in R$.

Note that in the present case $F(y)<F(x)$ follows for each $u \in U_P$ already from (10.1)-(10.3) if (10.3a) holds with "<".

An important consequence for the discretization of stochastic linear programs with recourse follows now from Theorem 2.2. In order to get the strict inequality $F(y)<F(x)$ by part (c) of Theorem 2.2, one requires that u is not affine-linear on $\mathrm{conv}\{z^i : i \in R\}$. Because of $u\binom{t}{\tilde z} = t+p(\tilde z)$, this means that p may not be affine-linear on

$\mathrm{conv}\{z_o^i : i \in R\}$, where $z_o^i = A_o^i x - b_o^i$. Since there are [27] convex cones $C_1, C_2, \ldots, C_\rho \subset \mathbb{R}^{m-1}$ and fixed vectors $f_1, f_2, \ldots, f_\rho \in \mathbb{R}^{m-1}$ such that $\bigcup_{\kappa=1}^{\rho} C_\kappa = \mathbb{R}^{m-1}$ and $p(\tilde{z}) = f_\kappa' \tilde{z}$ for all $\tilde{z} \in C_\kappa$, the above condition means that not all vectors $z_o^i, i \in R$, may lie in one and the same cone C_γ for some $1 \leq \gamma \leq \rho$. However, this yields an instruction for the discretization of a stochastic linear program with recourse at a certain point x (e.g. a certain iteration point)!

2.7. <u>Equivalent SD-conditions (12.1)-(12.3),(12.4a),(12.4b), (12.5) for the construction of feasible descent directions.</u> Collecting in $P_{A(\cdot)x-b(\cdot)}$ those m-vectors $z^i = A^i x - b^i, i \in R$, which are equal, we get

$$P_{A(\cdot)x-b(\cdot)} = \sum_{i \in S} \tilde{\alpha}_i \, \varepsilon_{z^i} \tag{11}$$

with a subset $S = S_x$ of R such that $z^i \neq z^j$ for all $i, j \in S$, $i \neq j$, and $\tilde{\alpha}_i = \sum_{z^t = z^i} \alpha_t$, hence $\tilde{\alpha}_i \geq \alpha_i$ and $\sum_{i \in S} \tilde{\alpha}_i = 1$. Consequently, in modification of (3), the SD-relations between $\mu = P_{A(\cdot)x-b(\cdot)}$ and $\nu = P_{A(\cdot)y-b(\cdot)}$ can also be given in the following form, where Π is replaced now by a stochastic matrix $T = (\tau_{ij})_{i \in S, j \in R}$:

(12.1) $\sum_{j \in R} \tau_{ij} = 1$, $\tau_{ij} \geq 0$, $i \in S$, $j \in R$,

(12.2) $\alpha_j = \sum_{i \in S} \tilde{\alpha}_i \tau_{ij}$, $j \in R$,

(12.3) $A^j y - b^j = \sum_{i \in S} \frac{\tilde{\alpha}_i \tau_{ij}}{\alpha_j} (A^i x - b^i)$, $j \in R$,

(12.4a) $K^j = \sum_{i \in S} \tilde{\beta}_{ij} \varepsilon_{z^i}$, $\tilde{\beta}_{ij} = \frac{\tilde{\alpha}_i \tau_{ij}}{\alpha_j}$, is not a one-point measure for at least one integer $j \in R$,

(12.4b) There is at least one $j \in R$ such that K^j is not a one-point measure and $\tau_{ij} > 0$ for all $i \in S$,

(12.5) $y \in D$.

Clearly, the conditions (5), $(3.3)_M$, $(3.4)_M$, $(3.3)_H$, $(9)_k$, $1 \le k \le m$, and (10) may be modified in the same way. It is easy to see that Theorem 2.2 - 2.4 and Lemma 2.2, 2.3 hold also with respect to these modified SD-conditions. Furthermore, there is a close relationship between the unknowns (y,Π) in (3) and (y,T) in (12). Indeed, a) let (y,Π) be a solution of (3.1)-(3.3), $\beta_{ij} = \frac{\alpha_i \pi_{ij}}{\alpha_j}$, and define $T = T(\Pi)$ by

$$\tau_{ij} = \frac{\sum_{z^t = z^i} \alpha_t \pi_{tj}}{\tilde{\alpha}_i} = \frac{\sum_{z^t = z^i} \alpha_t \pi_{tj}}{\sum_{z^t = z^i} \alpha_t}, \quad i \in S,\ j \in R. \tag{13.1}$$

Then (y,T) solves (12.1)-(12.3); moreover, since

$$\sum_{i \in R} \beta_{ij} \varepsilon_{z^i} = \sum_{i \in S} \Big(\sum_{z^t = z^i} \beta_{tj} \Big) \varepsilon_{z^i} = \sum_{i \in S} \Big(\frac{1}{\alpha_j} \sum_{z^t = z^i} \alpha_t \pi_{tj} \Big) \varepsilon_{z^i}$$

$$= \sum_{i \in S} \frac{\tilde{\alpha}_i \tau_{ij}}{\alpha_j} \varepsilon_{z^i} = \sum_{i \in S} \tilde{\beta}_{ij} \varepsilon_{z^i},$$

we find that (y,T) satisfies (12.4a), (12.4b) if (y,Π) satisfies (3.4a),(3.4b) respectively. b) Conversely, let (y,T) be a solution of (12.1)-(12.3) and define $\Pi = \Pi(T)$ by

$$\pi_{ij} = \tau_{sj}, \text{ where } s \in S \text{ is the unique index in } S \text{ such that } z^i = z^s, i,j \in R. \tag{13.2}$$

Then (y,Π) solves (3.1)-(3.3); furthermore, since again

$$\sum_{i \in R} \beta_{ij} \varepsilon_{z^i} = \sum_{i \in S} \frac{\tilde{\alpha}_i \tau_{ij}}{\alpha_j} \varepsilon_{z^i} = \sum_{i \in S} \tilde{\beta}_{ij} \varepsilon_{z^i}, \quad \tilde{\beta}_{ij} = \frac{\tilde{\alpha}_i \tau_{ij}}{\alpha_j},$$

we now have shown the equivalence of (3.4a), (3.4b) to (12.4a), (12.4b) respectively.

Note that $T(\Pi(T)) = T$, hence, each solution (y,T) of (a part of) (12) may be generated by a certain solution (y,Π) of (the corresponding part of) (3).

Example. Since $(y,\Pi) = (x,I)$, I = identity matrix, is a solution of (3.1)-(3.3), from above we obtain that $(y,T) = (x,T^o)$, with $T^o = T(I)$ given by

$$\tau_{ij}^o = \begin{cases} 0 \ , & z^i \neq z^j \\ \frac{\alpha_j}{\tilde{\alpha}_i} \ , & z^i = z^j \end{cases} \ , \ i \in S, j \in R, \qquad (14)$$

is a solution of (12.1)-(12.3).

Summarizing the considerations in section 2.7, we obtain the following lemma:

Lemma 2.4. The systems of relations (3) and (12) are equivalent in the sense that by means of (13) each solution (y,Π) of (3.1)-(3.3) resp. (3.1)-(3.4a)/(3.4b) generates a solution (y,T) of (12.1)-(12.3) resp. (12.1)-(12.4a)/(12.4b), and vice versa. Moreover, in this way each solution (y,T) of (a part of)(12) is generated by a certain solution (y,Π) of (the corresponding part of) (3).

Corresponding to (11), we may, of course, represent the distribution of $A(\omega)y-b(\omega)$ also by

$$P_{A(\cdot)y-b(\cdot)} = \sum_{j \in S_y} \alpha_{j,y} \, \varepsilon_{w^j}, \ w^j = A^j y - b^j, \qquad (11a)$$

where $w^j \neq w^t$ for $j,t \in S_y, j \neq t$, and $\alpha_{j,y} = \sum_{w^t = w^j} \alpha_t$. Consequently, the SD-conditions between $\mu = P_{A(\cdot)x-b(\cdot)}$ and $\nu = P_{A(\cdot)y-b(\cdot)}$ could be based in an equivalent way also on the distribution representations (11),(11a). However, since in contrary to the given vector x being e.g. the k-th iteration point of an algorithm, y is a yet unknown quantity, we have that the index set S_y is not known in advance. Hence, SD-conditions based on the distribution representations (11) and (11a) have no practical meaning.

3. Convex programs for solving (3.1)-(3.4a),(3.5)

3.1. A numerical criterion for (3.4a). For a given n-vector x let (y,Π) denote a solution of (3.1)-(3.3). According to section 2.7 for the transition probability measures $K^j, j \in R$, we have that

$$K^j = \sum_{i\in R} \beta_{ij}\, \varepsilon_{z^i} = \sum_{i \in S} \tilde{\beta}_{ij}\, \varepsilon_{z^i} \tag{15}$$

with $\sum_{i\in R} \beta_{ij} = \sum_{i\in S} \tilde{\beta}_{ij} = 1$, where $\beta_{ij} = \frac{\alpha_i \pi_{ij}}{\alpha_j}$, $i,j \in R$, and

$\tilde{\beta}_{ij} = \frac{\tilde{\alpha}_i \tau_{ij}}{\alpha_j}$, $\tilde{\alpha}_i = \sum_{z^t=z^i} \alpha_t$, $\tau_{ij} = \tau_{ij}(\Pi) = \frac{1}{\tilde{\alpha}_i} \sum_{z^t=z^i} \alpha_t \pi_{tj}$, $i \in S$,

$j \in R$, see (13); let $T=T(\Pi) = (\tau_{ij})$. In order to describe now numerically condition (3.4a) that K^j is not concentrated to one single point, for each $j \in R$ we introduce the functions

$$Q_j = Q_j(\Pi) = \tilde{Q}_j(T(\Pi)) = \sum_{i\in S} \tilde{\beta}_{ij}^2$$

and

$$M_j = M_j(\Pi) = \tilde{M}_j(T(\Pi)) = \sup_{i\in S} \tilde{\beta}_{ij}.$$

Since $0 \le \tilde{\beta}_{ij} \le 1$, $\sum_{i\in S} \tilde{\beta}_{ij} = 1$ and $\tilde{\beta}_{ij}^2 < \tilde{\beta}_{ij}$ for $0 < \tilde{\beta}_{ij} < 1$, it is $0 < Q_j, M_j \le 1$, $j \in R$; furthermore, $Q_j < 1$, $Q_j = 1$ holds if and only if $M_j < 1$, $M_j = 1$ respectively, where $Q_j = M_j = 1$ holds if and only if $\tilde{\beta}_{ij} = 1$ for a (unique) $i \in S$. Note that $Q_j = M_j = 1$ for $(y,\Pi) = (x,I)$. Because of (15) we have now this criterion.

Lemma 3.1. a) K^j is a one-point measure if and only if $Q_j = 1 (M_j = 1)$.
b) Condition (3.4a) holds if and only if $Q_j < 1 (M_j < 1)$ for at least one $j \in R$.

Given arbitrary, but fixed positive weights $\theta_j > 0$, $j \in R$, such that $\sum_{j\in R} \theta_j < +\infty$ and defining then the functions

$$Q(\Pi) = \tilde{Q}(T(\Pi)) = \sum_{j\in R} \theta_j Q_j \tag{16a}$$

and

$$M(\Pi) = \tilde{M}(T(\Pi)) = \sum_{j\in R} \theta_j M_j, \tag{16b}$$

we have, of course, also this criterion.

Corollary 3.1. Let $\theta_j>0$, $j\in R$, and $\sum_{j\in R}\theta_j<+\infty$. Then condition (3.4a) holds if and only if $Q<\sum_{j\in R}\theta_j$ $(M<\sum_{j\in R}\theta_j)$.

Note that always $0<Q\leq\sum_{j\in R}\theta_j$, $0<M\leq\sum_{j\in R}\theta_j$ and Q, M attain their maximal value $\sum_{j\in R}\theta_j$ e.g. for the trivial solution $(y,\Pi)=(x,I)$ of (3.1)-(3.3). The relationships between Q,M and the systems (3), (12) are shown now in the following theorem:

Theorem 3.1. For a given n-vector x let (y,Π) fulfill conditions (3.1)-(3.3). a) (y,Π) fulfills then also (3.4a) if and only if $Q(\Pi)<\sum_{j\in R}\theta_j$ $(M(\Pi)<\sum_{j\in R}\theta_j)$. b) If $Q(\Pi)=\sum_{j\in R}\theta_j$ $(M(\Pi)=\sum_{j\in R}\theta_j)$, then $P_{A(\cdot)y-b(\cdot)}=P_{A(\cdot)x-b(\cdot)}$ and therefore $F(y)=F(x)$ for all $u\in U_P$.

Proof. The first part (a) follows immediately from Corollary 3.1. b) The supposition $Q(\Pi)=\sum_{j\in R}\theta_j$ or $M(\Pi)=\sum_{j\in R}\theta_j$ means of course that $Q_j=M_j=1$ for every $j\in R$. Hence, referring to Lemma 3.1a, every K^j must be a one-point measure, i.e. for every $K^j, j\in R$, there exists a unique integer $i=q(j)\in S$ such that $K^j=\varepsilon_{z^{q(j)}}$ and therefore

$$\tilde{\beta}_{q(j)j}=\frac{1}{\alpha_j}\sum_{z^t=z^{q(j)}}\alpha_t\pi_{tj}=1$$

and

$$\tilde{\beta}_{ij}=\frac{1}{\alpha_j}\sum_{z^t=z^i}\alpha_t\pi_{tj}=0 \text{ for all } i\in S,\ i\neq q(j).$$

Consequently, since by assumption (y,Π) is a solution of (3.1)-(3.3), we obtain now for every $j\in R$ the equations

$$\text{a) } \sum_{z^t=z^{q(j)}}\alpha_t\pi_{tj}=\alpha_j, \tag{17a}$$

$$\text{b) } \pi_{tj}=0 \text{ if } t\in R,\ z^t\neq z^{q(j)} \tag{17b}$$

$$\text{c) } w^j=A^jy-b^j=\sum_{i\in R}\beta_{ij}z^i=\sum_{i\in S}\tilde{\beta}_{ij}z^i=z^{q(j)}. \tag{17c}$$

For every fixed $t\in R$ from (3.1) and (17b) follows that

$\sum^j_{z^{q(j)}=z^t}\pi_{tj}=1$ (For sake of uniqueness the index "j" in Σ_j denotes the summation index), hence for $t=i \in S$ we find that $\sum_{j:\, q(j)=i} \pi_{ij} = 1$. This implies of course that the index set $J_i=\{j \in R:\ q(j)=1\}$ is non-empty for every $i \in S$. Moreover, $\{J_i : i \in S\}$ is a disjoint partition of R, and for $i \in S$ it is, see (17a),

$$\sum_{j\in J_i} \alpha_j = \sum_{j\in J_i} \Big(\sum_{t:\, z^t=z^{q(j)}} \alpha_t \pi_{tj} \Big) = \sum_{j\in J_i} \Big(\sum_{t:\, z^t=z^i} \alpha_t \pi_{tj} \Big)$$

$$= \sum_{t:\, z^t=z^i} \alpha_t \Big(\sum_{j\in J_i} \pi_{tj} \Big) = \sum_{t:\, z^t=z^i} \alpha_t \Big(\sum_{j:\, q(j)=i} \pi_{tj} \Big)$$

$$= \sum_{t:\, z^t=z^i} \alpha_t \Big(\sum_{j:\, z^{q(j)}=z^i} \pi_{tj} \Big) = \sum_{t:\, z^t=z^i} \alpha_t \Big(\sum_{j:\, z^{q(j)}=z^t} \pi_{tj} \Big)$$

$$= \sum_{t:\, z^t=z^i} \alpha_t = \tilde{\alpha}_i,$$

since $\sum_{j:\, z^{q(j)}=z^t} \pi_{tj} = 1$ for every $t \in R$. This representation of $\tilde{\alpha}_i$ together with (17c) finally yields now

$$\int f \, dP_{A(\cdot)y-b(\cdot)} = \sum_{j\in R} \alpha_j f(w^j) = \sum_{i\in S} \Big(\sum_{j\in J_i} \alpha_j f(w^j) \Big)$$

$$= \sum_{i\in S} \Big(\sum_{j\in J_i} \alpha_j f(z^{q(j)}) \Big) = \sum_{i\in S} \Big(\sum_{j\in J_i} \alpha_j f(z^i) \Big)$$

$$= \sum_{i\in S} \Big(\sum_{j\in J_i} \alpha_j \Big) f(z^i) = \sum_{i\in S} \tilde{\alpha}_i f(z^i)$$

$$= \sum_{i\in R} \alpha_i f(z^i) = \int f \, dP_{A(\cdot)x-b(\cdot)}$$

for every bounded function f on $\mathbb{R}^m$, hence $P_{A(\cdot)y-b(\cdot)} = P_{A(\cdot)x-b(\cdot)}$.

3.2. <u>Convex programs $(P_{x,D})$ for solving (3.1)-(3.4a),(3.5).</u> For a given n-vector x let (y,Π) be a solution of the linear relations (3.1)-(3.3). Since (y,Π) fulfills then also (3.4a) if and only if $Q(\Pi)(M(\Pi))$ is less than its maximal value $\sum_{j\in R} \theta_j$, solutions of (3.1)-(3.4a) may be constructed by minimizing $Q(\Pi)$ or $M(\Pi)$ subject to the constraints (3.1)-(3.3), i.e. by solving one of the following finite or infinite convex minimization problems

(P_x)

minimize Q s.t. (3.1)-(3.3) (P_x^Q)

minimize M s.t. (3.1)-(3.3). (P_x^M)

Obviously, (P_x^Q), (P_x^M) is a quadratic respectively linear program. Both programs have always the feasible solution $(y,\Pi)=(x,I)$, where $Q(I)=\tilde{Q}(T^o)=\sum_{j\in R}\theta_j$ and $M(I)=\tilde{M}(T^o)=\sum_{j\in R}\theta_j$, see (14). Since $Q(\Pi)=\tilde{Q}(T(\Pi))$ and $\tilde{Q}(T)=\sum_{j\in R}\theta_j\sum_{i\in S}(\frac{\tilde{\alpha}_i\tau_{ij}}{\alpha_j})^2$ is a strictly convex function of T, the T-transform $T(\Pi^*)$ of the second component Π^* of an optimal solution (y^*,Π^*) of (P_x^Q) is uniquely determined, i.e. if (y^{**},Π^{**}) is a further optimal solution of (P_x^Q), then $T(\Pi^*)=T(\Pi^{**})$. From Theorem 3.1 we obtain this next corollary:

Corollary 3.2. For a given $x\in D$ let (y^*,Π^*) be an optimal solution of (P_x^Q), (P_x^M) and define $Q^*=Q(\Pi^*)$, $M^*=M(\Pi^*)$ respectively. a) If $Q^*<\sum_{j\in R}\theta_j$ $(M^*<\sum_{j\in R}\theta_j)$, then (y^*,Π^*) fulfills (3.1)-(3.4a). b) If $Q^*=\sum_{j\in R}\theta_j$ $(M^*=\sum_{j\in R}\theta_j)$, then (3.1)-(3.4a) has no solution, but it is $P_{A(\cdot)x^*-b(\cdot)}=P_{A(\cdot)x-b(\cdot)}$.

In our basic optimization problem (SOP) we have the constraint "$y\in D$", where D is a certain convex subset of $\mathbb{R}^n$. If (y,Π), $y\neq x$, is a feasible solution of (P_x) with $x\in D$, but we don't know whether x lies also in the interior of D, then, in general, we have no guarantee that $h=y-x$ is a feasible descent direction of F at x with respect to D. Hence, in order to get directly feasible directions of decrease of F at $x\in D$ with respect to D, we may still add to (P_x) to constraint (3.5), which yields then the programs

minimize Q s.t. (3.1)-(3.3), (3.5) $(P_{x,D}^Q)$

minimize M s.t. (3.1)-(3.3), (3.5). $(P_{x,D}^M)$

3.3. Further convex programs for solving (3.1)-(3.4a),(3.5). Besides Q_j and M_j, there are many other measures of uncertainty [10] which may be used to describe that K^j is not concentrated to one point. We still mention two important ones.

3.3.1. Entropy maximization. The entropy H_j of K_j is defined [23] by

$$H_j = H_j(\Pi) = \tilde{H}_j(\tau(\Pi)) = - \sum_{i \in S} \tilde{\beta}_{ij} \log \tilde{\beta}_{ij}.$$

It is $H_j \geq 0$, where $H_j=0$ holds if and only if K^j is a one-point measure; note that $H_j \leq \log|S|$. Therefore, corresponding to the criterions in Corollary 3.1, (3.4a) holds if and only if $H(\Pi)>0$, where

$$H(\Pi) = \sum_{j \in R} \theta_j H_j, \tag{16c}$$

with positive weights $\theta_j>0$, $j \in R$, such that $\sum_{j \in R} \theta_j<+\infty$, and instead of $(P^Q_{x,D})$ or $(P^M_{x,D})$ we may also employ the concave (entropy-) maximization problem

$$\text{minimize } -H \text{ s.t. } (3.1)\text{-}(3.3),(3.5). \tag{$P^H_{x,D}$}$$

Note. a) It is $H(\Pi)=\tilde{H}(\tau(\Pi))$, where $H(\tau)= \sum_{j \in R} \theta_j \tilde{H}_j(\tau)$ is strictly concave in τ. b) Entropy maximization is often studied in statistics and engineering, see e.g. [44],[49].

3.3.2. Variance maximization. Not quite satisfactory - from the practical point of view - is that, working with Q_j, M_j, H_j or similar measures of uncertainty based directly on the representation $\sum_{i \in S} \tilde{\beta}_{ij} \varepsilon_{z^i}$ of K^j, one has first to determine explicitly the index set $S=S_x$, see section 2.7. This can be omitted if we consider the sum of the variances of the components $z_1,\dots,z_m$ of z interpreted as an m-random vector having distribution K^j, hence

$$V_j = V_j(\Pi) = \operatorname{tr} \operatorname{cov}(z|K^j)$$
$$= \sum_{i \in R} \beta_{ij} ||z^i-w^j||^2_E = \sum_{i \in R} \beta_{ij} ||z^i- \sum_{i \in R} \beta_{ij} z^i||^2_E =$$

$$= \sum_{i\in R} \beta_{ij}||z^i||_E^2 - ||\sum_{i\in R} \beta_{ij} z^i||_E^2,$$

where $||\cdot||_E$ denotes the Euclidean norm. Of course, replacing β_{ij} by $\tilde{\beta}_{ij} = \frac{\tilde{\alpha}_i \tau_{ij}}{\alpha_j}$, we find $V_j(\Pi) = \tilde{V}_j(T(\Pi))$ with $\tilde{V}_j(T) = \sum_{i\in S} \tilde{\beta}_{ij}||z^i||_E^2 - ||\sum_{i\in S} \tilde{\beta}_{ij} z^i||_E^2$, where V_j, $\tilde{V}_j$ is concave in Π resp. in T. Moreover, K^j is not concentrated to one point if and only if $V_j>0$. Therefore, (3.4a) holds if and only if $V(\Pi)>0$, where $V(\Pi)$ is the concave quadratic function

$$V(\Pi) = \sum_{j\in R} \theta_j V_j(\Pi) \tag{16d}$$

with positive weights $\theta_j>0, j\in R$, such that $\sum_{j\in R} \theta_j<+\infty$. Instead of $(P^Q_{x,D})$, $(P^M_{x,D})$ or $(P^H_{x,D})$ we can then also use the concave variance maximization problem

$$\text{minimize } -V \text{ s.t. (3.1)-(3.3),(3.5).} \qquad (P^V_{x,D})$$

Obviously, the propositions proved in this section hold also under some slight modifications for $(P^H_{x,D})$ and $(P^V_{x,D})$. In the following, by $(P_{x,D})$ we understand any one of the four programs considered above.

4. Stationary points (efficient solutions) of (SOP)

For a given $x \in D$, the descent direction - finding procedure based on the program $(P_{x,D})$ can only fail completely if for each feasible solution (y,Π) of $(P_{x,D})$ it holds $A^j y = A^j x$ for all $j \in R$. Indeed, in this situation we either have $y=x$, or F is constant on the whole line through the points x and y, hence, $h=y-x$ is not a descent direction and the procedure fails to find a descent direction of F at x. This observation suggests the following definition of stationarity.

Definition 4.1. A point $x \in D$ is called D-stationary (stationary relative to D) or an efficient solution of (SOP) if $(P_{x,D})$ has only feasible solutions (y,Π) such that $A^j y = A^j x$ for all $j \in R$, i.e. $A(\omega)\cdot(y-x)=0$ w.p.1.

4.1. Necessary optimality conditions without using derivatives.

If $x \in D$ is D-stationary, then the procedure for constructing feasible descent directions $h=y-x$ based on $(P_{x,D})$ fails at x. However, since in an arbitrary optimal solution x^* of the basic optimization problem (SOP) there cannot exist any feasible descent direction, stationary points are candidates for optimal solutions of (SOP), as is shown in the following lemma:

Lemma 4.1. (Necessary optimality condition without subgradient of F). Suppose that for every $x \in D$ and every feasible solution (y,Π) of $(P_{x,D})$ the objective function F of (SOP) is constant on the line segment $\overline{xy}$ if and only if $A^j y = A^j x$ for all $j \in R$. If x^* is any optimal solution of (SOP), then x^* is also D-stationary.

Proof. Assume that x^* is not D-stationary. Then, by Definition 4.1 there exists a feasible solution (y,Π) of $(P_{x,D})$ such that $A^j y \neq A^j x^*$ for at least one $j \in R$, where $y \in D$. This yields $y \neq x^*$, and from Theorem 2.2 it follows that $F(y) \leq F(x^*)$; furthermore, F is not constant on $\overline{x^* y}$, which is now a contradiction to the optimality of x^*. Hence, x^* is also D-stationary.

Note. The assumption in Lemma 4.1 concerning F is fulfilled e.g. if u is a strictly convex loss function.

Since the set S_D of D-stationary points contains, under weak assumptions, the optimal solutions x^* of (SOP), in the following we study the stationarity concept in more detail.

4.2. <u>The form of a feasible solution (y,Π) of $(P_{x,D})$ for a D-stationary point x.</u> Given a D-stationary point x, we are interested first in the form of the second component Π in a feasible solution (y,Π) of $(P_{x,D})$. According to Definition 4.1, for the first component y of (y,Π) we have that $A^j y = A^j x$ for every $j \in R$; note that this implies $y=x$, provided that the $|R| x n$ matrix

$$\mathbb{A} = \begin{pmatrix} \vdots \\ A^j \\ \vdots \end{pmatrix}_{j \in R} \tag{18}$$

contains a regular nxn submatrix. The wanted form of Π is described as follows:

<u>Theorem 4.1.</u> For given $x \in \mathbb{R}^n$ let (y,Π) fulfill conditions (3.1)-(3.3). If $A^j y = A^j x$ for all $j \in R$, and $Z_x = \{z^i : i \in R\}$ is a compact subset of $\mathbb{R}^m$, then $T(\Pi) = T^0$, where $T(\Pi)$, T^0 are defined by (13.1),(14) respectively.

Proof. Because of $w^j = A^j y - b^j = A^j x - b^j = z^j$, from (3.3) follows for each $j \in R$ that

$$z^j = w^j = \sum_{j \in R} \frac{\alpha_i \pi_{ij}}{\alpha_j} z^i = \sum_{i \in S} \tilde{\beta}_{ij} z^i, \tag{19}$$

where $\tilde{\beta}_{ij} = \frac{\tilde{\alpha}_i \tau_{ij}}{\alpha_j}$ and $\tau_{ij} = \frac{1}{\tilde{\alpha}_i} \sum_{z^t = z^i} \alpha_t \pi_{tj}$, see (13.1). Let $Z_1 := Z_x = \{z^i : i \in S\}$ and consider the closed convex hull $C_1 = \overline{\text{conv}}\, Z_1 = \overline{\text{conv}\, Z_1}$ of Z_1. The assumption that Z_1 is compact implies that also C_1 is compact, hence $C_1 = \overline{\text{conv}}\, Z_1^e$, where Z_1^e is the set of extreme points of C_1 and $Z_1^e \subset Z_1$, see e.g. [12]. Because of (19) we also must have

$$z^j = \sum_{i \in S} \tilde{\beta}_{ij} z^i \text{ for each } z^j \in Z_1^e. \tag{19a}$$

However, since $\sum_{i \in S} \tilde{\beta}_{ij} z^i \in C_1$ and $z^j \in Z_1^e$ is an extreme point of C_1, equation (19a) can hold if and only if for each $z^j \in Z_1^e$ it is

$$\frac{\tilde{\alpha}_i \tau_{ij}}{\alpha_j} = \tilde{\beta}_{ij} = \begin{cases} 1, & \text{if } z^i = z^j, i \in S \\ 0, & \text{if } z^i \neq z^j, i \in S \end{cases},$$

hence

$$\tau_{ij} = \begin{cases} \dfrac{\alpha_j}{\tilde{\alpha}_i}, & \text{if } z^i = z^j, i \in S \\ 0, & \text{if } z^i \neq z^j, i \in S \end{cases} \quad \text{for each } z^j \in Z_1^e. \tag{20}$$

For any fixed index $i \in S$ such that $z^i \in Z_1^e$ from (20) follows that

$$\sum_{\substack{j \\ z^j \in Z_1^e}} \tau_{ij} = \sum_{\substack{j \\ z^j \in Z_1^e \\ z^j \neq z^i}} \tau_{ij} + \sum_{\substack{j \\ z^j \in Z_1^e \\ z^j = z^i}} \tau_{ij}$$

$$= \sum_{\substack{j \\ z^j \in Z_1^e \\ z^j = z^i}} \tau_{ij} = \sum_{\substack{j \\ z^j = z^i}} \frac{\alpha_j}{\tilde{\alpha}_i} = 1,$$

hence, because of $\sum_{j \in R} \tau_{ij} = 1$ we must have

$$\tau_{ij} = 0, \text{ if } z^i \in Z_1^e, \ i \in S \text{ and } z^j \notin Z_1^e. \tag{21}$$

Our relations (19) and (21) yield now for each index $j \in R$ with $z^j \notin Z_1^e$ that

$$z^j = \sum_{i \in S} \tilde{\beta}_{ij} z^i = \sum_{\substack{i \in S \\ z^i \in Z_1^e}} \frac{\tilde{\alpha}_i \tau_{ij}}{\alpha_j} z^i + \sum_{\substack{i \in S \\ z^i \notin Z_1^e}} \frac{\tilde{\alpha}_i \tau_{ij}}{\alpha_j} z^i$$

$$= \sum_{\substack{i \in S \\ z^i \notin Z_1^e}} \frac{\tilde{\alpha}_i \tau_{ij}}{\alpha_j} z^i$$

as well as

$$1 = \sum_{i \in S} \tilde{\beta}_{ij} = \sum_{\substack{i \in S \\ z^i \in Z_1^e}} \frac{\tilde{\alpha}_i \tau_{ij}}{\alpha_j} + \sum_{\substack{i \in S \\ z^i \notin Z_1^e}} \frac{\tilde{\alpha}_i \tau_{ij}}{\alpha_j}$$

$$= \sum_{\substack{i \in S \\ z^i \notin Z_1^e}} \frac{\tilde{\alpha}_i \tau_{ij}}{\alpha_j},$$

hence, it is

$$z^j = \sum_{\substack{i \in S \\ z^i \notin Z_1^e}} \tilde{\beta}_{ij} z^i, \ 1 = \sum_{\substack{i \in S \\ z^i \notin Z_1^e}} \tilde{\beta}_{ij} \text{ for each } z^j \notin Z_1^e. \tag{22}$$

Obviously, (22) is exactly of the same type as (19), however, (22) involves only the vectors z^j from $Z_2=\{z^j : z^j \notin Z_1^e\}$, where $Z_2 \subset Z_1$ and $Z_2 \neq Z_1$, since $Z_1^e \neq \emptyset$. Replacing now Z_1 by Z_2 we may therefore repeat the above procedure, which then generates Z_2^e and a next submatrix of the type (20), but being completely separated from the first one. Repeating this procedure again and again, we finally und up with the asserted matrix $T(\Pi)=T^o$.

It is easy to see that with respect to the relation system (12), the above theorem can also be formulated this way.

Corollary 4.1. For given $x \in \mathbb{R}^n$ let (y,T) fulfill conditions (12.1)-(12.3). If $A^j y=A^j x$ for all $j \in R$ and Z_x is compact, then $T=T^o$.

Note. If R is finite, then Z_x is compact.

4.3. The reversal of 4.2. Conversely, we are now interested in the form of the first component y in a feasible solution (y,Π) of $(P_{x,D})$ such that $T(\Pi)=T^o$, where $T(\Pi), T^o$ are defined by (13.1), (14) respectively.

Theorem 4.2. For a given $x \in \mathbb{R}^n$ let (y,Π) fulfill (3.1)-(3.3). If $T(\Pi)=T^o$, then $A^j y = A^j x$ for every $j \in R$. If $\mathbf{A}$ (see (18)) contains a regular nxn submatrix, then $T(\Pi)=T^o$ implies $y=x$.

Proof. According to the assumptions it is, see (13.1), (14),

$$\tau_{ij} = \tau_{ij}^o = \begin{cases} 0\ , & z^i \neq z^j \\ \dfrac{\alpha_j}{\tilde{\alpha}_i}, & z^i = z^j \end{cases}, \quad i \in S, j \in R.$$

From section 2 and (3.3) follows then that

$$A^j y - b^j = \sum_{i \in R} \frac{\alpha_i \pi_{ij}}{\alpha_j} z^i = \sum_{i \in S} \frac{\tilde{\alpha}_i \tau_{ij}}{\alpha_j} z^i$$

$$= \sum_{\substack{i \in S \\ z^i = z^j}} \frac{\tilde{\alpha}_i \tau_{ij}}{\alpha_j} z^i + \sum_{\substack{i \in S \\ z^i \neq z^j}} \frac{\tilde{\alpha}_i \tau_{ij}}{\alpha_j} z^i = \sum_{\substack{i \in S \\ z^i = z^j}} \frac{\tilde{\alpha}_i \tau_{ij}^o}{\alpha_j} z^i .$$

Since to each $j \in R$ there is a unique $i = i_j \in S$ such that $z^j = z^{i_j}$, we obtain

$$A^j y - b^j = \frac{\tilde{\alpha}_{i_j} \tau_{i_j j}^o}{\alpha_j} z^{i_j} = z^{i_j} = z^j = A^j x - b^j ,$$

hence, $A^j y = A^j x$ for all $j \in R$.

With respect to (12), the above result can also be given in this form.

Corollary 4.2. For given $x \in \mathbb{R}^n$ let (y,T) fulfill (12.1)-(12.3). If $T=T^o$, then $A^j y = A^j x$ for every $j \in R$.

From Theorem 4.1 and 4.2 we still obtain this corollary:

Corollary 4.3. For a given n-vector x let $Z_x = \{z^i : i \in R\}$ be compact. a) If (y,Π) fulfills (3.1)-(3.3), then $A^j y = A^j x$ for every $j \in R$ if and only if $T(\Pi)=T^o$, where $T(\Pi)$, T^o are defined by (13.1), (14) respectively. b) If (y,T) fulfills (12.1)-(12.3), then $A^j y = A^j x$ for every $j \in R$ if and only if $T=T^o$.

4.4. Characterizations of efficient solutions. According to Corollary 4.3, stationary points may be described as follows.

Theorem 4.3. a) A point $x \in D$ with compact Z_x is stationary relative to D if and only if $(P_{x,D})$ has only feasible solutions (y,Π) such that $A^j y = A^j x$ for every $j \in R$ and $T(\Pi)=T^o$. b) A point $x \in D$ with compact Z_x is stationary relative to D if and only if $(P^Q_{x,D})$ (or $(P^H_{x,D})$) has an optimal solution (y^*,Π^*) such that $A^j y^* = A^j x$ for every $j \in R$ and $T(\Pi^*)=T^o$.

Proof. a) According to Definition 4.1 a point $x \in D$ is stationary relative to D if and only if for each feasible solution (y,Π) of $(P_{x,D})$ it is $A^j y = A^j x$ for every $j \in R$, hence, because of Corollary 4.3, we must also simultaneously have $T(\Pi)=T_o$. b) If $x \in D$ is stationary, then for all feasible solutions (y,Π) is holds $A^j y = A^j x$ for every $j \in R$ and $T(\Pi)=T^o$, hence $Q(\Pi)=\tilde{Q}(T(\Pi))=\tilde{Q}(T^o)=\sum_{j \in R} \theta_j$. This means that for stationary $x \in D$ all feasible solutions (y,Π) of $(P_{x,D})$ are also optimal solutions, where $A^j y = A^j x$, $j \in R$ and $T(\Pi)=T^o$. Conversely, let (y^*,Π^*) be an optimal solution of $(P^Q_{x,D})$ such that $A^j y^* = A^j x$, $j \in R$ and $T(\Pi^*)=T^o$, hence $Q(\Pi^*)=\tilde{Q}(T^o)=\sum_{j \in R} \theta_j$. For an arbitrary feasible solution (y,Π) of $(P_{x,D})$ we then obtain $\sum_{j \in R} \theta_j \geq Q(\Pi) \geq Q(\Pi^*)=\sum_{j \in R} \theta_j$, hence (y,Π) is also optimal in $(P^Q_{x,D})$. Since $T(\Pi^*)$ is uniquely determined, see section 3, we must have $T(\Pi)=T(\Pi^*)=T^o$ and therefore $A^j y = A^j x, j \in R$, see Corollary 4.3, hence x is stationary. The assertion concerning $(P^H_{x,D})$ follows in a quite similar way.

Note. Because of Corollary 4.3 the second part of Theorem 4.3 can be formulated this way: A point $x \in D$ with compact Z_x is stationary relative to D if and only if $(P^Q_{x,D})$ has an optimal solution (y^*,Π^*) such that $T(\Pi^*)=T^o$ (or $A^j y^* = A^j x$ for every $j \in R$).

For a given sequence (θ_j) of positive numbers $\theta_j>0$, $j \in R$, such that $\sum_{j \in R} \theta_j ||A^j x||^2_E < +\infty$ for all $x \in D$, consider finally the program minimize - N s.t. (3.1)-(3.3),(3.5), $(P^N_{x,D})$

where, for given $x \in D$, the objective function N is defined by

$$N = N(y) = \sum_{j \in R} \theta_j ||A^j y - A^j x||_E^2 = (y-x)'(\sum_{j \in R} \theta_j A^{j\prime} A^j)(y-x). \quad (16e)$$

If (row)rank $\mathbb{A}$=n, i.e. if $\mathbb{A}$ contains a regular nxn submatrix, then N(y) may be defined by

$$N(y) = ||y-x||_E^2. \quad (16e)'$$

It is easy to see that Theorem 4.3 can then be supplemented as follows.

Theorem 4.3.1. A point $x \in D$ is D-stationary if and only if the supremum N^* of $(P^N_{x,D})$ is equal to zero. A point $x \in D$ with compact Z_x is D-stationary if and only if $(P^N_{x,D})$ has an optimal solution (y^*, Π^*) such that $N(y^*)=0$, i.e. $A^j y^* = A^j x$ for every $j \in R$, and $T(\Pi)=T^0$.

Unfortunately, since N is convex, $(P^N_{x,D})$ is not a concave maximization problem, hence the Kuhn-Tucker conditions are only nesessary, see [25].

5. Optimal solutions of $(P_{x,D})$, $(\tilde{P}_{x,D})$

For a given fixed element x of D, let $(P_{x,D})$ denote any one of the auxiliary programs $(P^C_{x,D})$, C=Q,M,H,V,N, considered in the preceding sections 3,4 for the construction of solutions (y,Π) of (3.1)-(3.3),(3.4a)/(3.4b),(3.5) or for the characterization of D-stationary points. For simplification, suppose that $(A(\omega),b(\omega))$ has a finite spectrum, i.e. $R=\{1,2,...,r\}$ for some integer r. Hence, S is also finite and Z_x is compact.

5.1. Existence of optimal solutions. Concerning the existence of optimal solutions of $(P_{x,D})$, we have this lemma:

Lemma 5.1. Let $x \in D$. a) If D is compact, then $(P_{x,D})$ has an optimal solution (y^*,Π^*). b) If D is a convex polyhedral set, then $(P^C_{x,D})$ has an optimal solution (y^*,Π^*) for every C=Q, M and V.

Proof. a) If D is compact, then the joint set of feasible solutions of all five programs $(P_{x,D})$ is non-empty and compact, which yields the assertion, since obviously the objective functions C=Q,M,H,V and N are continuous. b) If D is convex and polyhedral, then $(P^Q_{x,D})$, $(P^V_{x,D})$ resp. $(P^M_{x,D})$ are convex quadratic resp. linear minimization problems having objective functions Q,V and M, which are bounded from below. Hence, the existence of an optimal solution (y^*,Π^*) of $(P_{x,D})$ follows, in the present case, from the existence theorems of linear and quadratic programming, see e.g. [7].

Note. If rank $\mathbb{A}=n$, where $\mathbb{A}'=(A^{1'},A^{2'},...,A^{r'})$, then Lemma 5.1a holds also under the weaker assumption that D is closed.

5.2. Optimality conditions. Suppose now that D is defined by

$$D = \{x \in \mathbb{R}^n: g_k(x) \leq 0, k=1,2,...,\kappa\} \quad (23)$$

where $g_1,...,g_\kappa$ are given differentiable, convex functions. In order to establish the local optimality conditions for $(P_{x,D})$, we consider here only the differentiable objective functions C=Q,-H,-V and -N (we are always minimizing). Since by assumption R is finite, in C we may take the weights $\theta_j=1$ for every $j \in R$. Since $C=C(y,\Pi)=\tilde{C}(y,T(\Pi))$ and due to the transformations $\Pi \to T(\Pi)$, $T \to \Pi(T)$, see section 2.7 and Lemma 2.4, the programs $(P_{x,D})$ may be replaced by the equivalent programs

$$\text{minimize } \tilde{C}(y,T) \text{ s.t. } (12.1)\text{-}(12.3),(12.5), \qquad (\tilde{P}_{x,D})$$

$\tilde{C}=\tilde{Q}(T)$, $-\tilde{H}(T)$, $-\tilde{V}(T)$ or $\tilde{C}=-N(y)$, which have the Lagrangians

$$\tilde{L} = \tilde{L}(y,T,\mu_i,\lambda_j,\gamma_j,\rho_k, i \in S, j \in R, 1 \le k \le \kappa)$$

$$= \tilde{C}(y,T) + \sum_{i \in S} \mu_i (\sum_{j \in R} \tau_{ij} - 1) + \sum_{j \in R} \lambda_j (\sum_{i \in S} \tilde{\alpha}_i \tau_{ij} - \alpha_j)$$

$$+ \sum_{j \in R} \gamma_j' (\sum_{i \in S} \frac{\tilde{\alpha}_i \tau_{ij}}{\alpha_j} z^i - (A^j y - b^j)) + \sum_{k=1}^{\kappa} \rho_k g_k(y),$$

where $\mu_i \in \mathbb{R}$, $\lambda_j \in \mathbb{R}$, $\gamma_j \in \mathbb{R}^m$ and $\rho_k \ge 0$ for $i \in S, j \in R$ and $1 \le k \le \kappa$ are the Lagrange multipliers relative to the constraints (12.1)-(12.3), (12.5). The partial derivatives of $\tilde{L}$ with respect to y and τ_{ij}, $i \in S, j \in R$, are given by

$$\frac{\partial \tilde{L}}{\partial y} = \begin{cases} - \sum_{j \in R} A^{j\prime} \gamma_j + \sum_{k=1}^{\kappa} \rho_k \nabla g_k(y), \text{ for } \tilde{C}=\tilde{Q},-\tilde{H},-\tilde{V} \\ \\ - 2N_o(y-x) - \sum_{j \in R} A^{j\prime} \gamma_j + \sum_{k=1}^{\kappa} \rho_k \nabla g_k(y), \text{ for } \tilde{C}=-N, \end{cases} \tag{24}$$

where $N_o = \sum_{j \in R} A^{j\prime} A^j$ ($N_o=I$, resp.), and

$$\frac{\partial \tilde{L}}{\partial \tau_{ij}} = \begin{cases} \frac{\tilde{\alpha}_i}{\alpha_j} \frac{\partial \tilde{C}_j}{\partial \beta_{ij}} + \mu_i + \lambda_j \tilde{\alpha}_i + \frac{\tilde{\alpha}_i}{\alpha_j} \gamma_j' z^i, \text{ for } \tilde{C}=\tilde{Q},-\tilde{H},-\tilde{V} \\ \\ \mu_i + \lambda_j \tilde{\alpha}_i + \frac{\tilde{\alpha}_i}{\alpha_j} \gamma_j' z^i, \text{ for } \tilde{C}=-N \end{cases} \tag{25}$$

with

$$\frac{\partial \tilde{C}_j}{\partial \beta_{ij}} = \begin{cases} 2\ \tilde{\beta}_{ij}, & \text{if } \tilde{C}=\tilde{Q} \\ 1 + \log \tilde{\beta}_{ij}, & \text{if } \tilde{C}=-\tilde{H} \\ -z^{i\prime}z^i + 2\ z^{i\prime} \sum_{t\in S} \tilde{\beta}_{tj}\ z^t, & \text{if } \tilde{C}=-\tilde{V}, \end{cases} \tag{26}$$

where $\sum_{i\in S} \tilde{\beta}_{ij}\ z^i = w^j$ if (12.3) holds. For a convex polyhedral set D, i.e. if $g(y)=(g_1(y),\dots,g_\kappa(y))'$ is given by

$$g(y) = Gy - g \tag{27}$$

with a fixed $\kappa x(n+1)$ matrix (G,g), then

$$\sum_{k=1}^{\kappa} \rho_k\ \nabla g_k(y) = G'\rho,\ \rho=(\rho_1,\dots,\rho_\kappa)'. \tag{28}$$

For an optimal solution (y^*,T^*) of $(\tilde{P}_{x,D})$ we have [7] the local Kuhn-Tucker conditions

$$\frac{\partial \tilde{L}}{\partial y} = 0 \tag{29.1}$$

$$\frac{\partial \tilde{L}}{\partial \tau_{ij}} \geq 0,\ \tau_{ij}\ \frac{\partial \tilde{L}}{\partial \tau_{ij}} = 0,\ \tau_{ij} \geq 0 \text{ for all } i \in S, j \in R \tag{29.2}$$

$$g_k \leq 0,\ \rho_k\ g_k = 0,\ \rho_k \geq 0,\ k=1,\dots,\kappa \tag{29.3}$$

and of course the remaining conditions

$$(12.1) - (12.3). \tag{29.4}$$

From optimization theory [7], [25] we may derive the next result:

Lemma 5.2. Suppose that $(\tilde{P}_{x,D})$ has a feasible solution (y,T) such that $g_k(y)<0$ for every non-affine linear g_k, $k=1,\dots,\kappa$. Then (y^*,T^*) is an optimal solution of $(\tilde{P}_{x,D})$ with $\tilde{C}=\tilde{Q}$ or $\tilde{C}=-\tilde{V}$ if and only if there exist Lagrange multipliers μ_i^*, λ_j^*, γ_j^*, $i \in S, j \in R$, and ρ^* such that the tuple $((y^*,T^*), (\mu_i^*), (\lambda_j^*), (\gamma_j^*), \rho^*)$ fulfills condition (29).

Proof. The programs $(\tilde{P}^Q_{x,D})$, $(\tilde{P}^V_{x,D})$ are convex programs having a differentiable objective function and differentiable functions in the constraints. The assertion follows then from the theory of convex optimization.

5.3. Characterization of stationary points by optimality conditions.

5.3.1. Characterization by means of $(\tilde{P}^{Q}_{x,D})$. Stating first Theorem 4.3 in the variables (y,T), we obtain the following corollary:

Corollary 5.1. A point $x \in D$ is D-stationary if and only if $(\tilde{P}^{Q}_{x,D})$ has an optimal solution (y^*,T^*) such that $T^*=T^o$ and $A^j y^* = A^j x$ for every $j \in R$.

For simplification we now assume that D is the convex polyhedron given by (23) and (27). Since certainly every tuple (y,T^o) with $A^j y = A^j x$, $j \in R$, fulfills (12.1)-(12.3), from Lemma 5.1, 5.2 and Corollary 5.1 we get this characterization:

Theorem 5.1. Let D be given by (23) and (27). Then $x \in D$ is D-stationary if and only if there exist an n-vector y, numbers μ_i, λ_j, ρ_k and m-vector γ_j, $i \in S$, $j \in R$, $1 \le k \le \kappa$, such that the tuple $((y,T^o), (\mu_i), (\lambda_j), (\gamma_j), \rho)$ satisfies (29.1)-(29.3) with $\tilde{C}=\tilde{Q}$, where $A^j y = A^j x$ for every $j \in R$.

Explicitly, for a D-stationary point x we have the following necessary and sufficient conditions (30): First of all it must hold

$$A^j y = A^j x \text{ for every } j \in R, \tag{30.1}$$

which simply means

$$y = x, \tag{30.1)'$$

provided that rank $\mathbb{A}=n$. From (24),(28) and (29.1) we obtain

$$0 = - \sum_{j \in R} A^{j'} \gamma_j + G'\rho. \tag{30.2}$$

Condition (29.3) together with $x \in D$ means that

$$Gx \le g, \tag{30.3a}$$

$$Gy \le g,\ \rho'(Gy-g) = 0,\ \rho \ge 0. \tag{30.3b}$$

Finally, (25),(26),(29.2) and the definition (14) of T^o yield the conditions

$$2\frac{\tilde{\alpha}_i}{\alpha_j} + \mu_i + \lambda_j\tilde{\alpha}_i + \frac{\tilde{\alpha}_i}{\alpha_j}\gamma_j{}'z^i = 0, \text{ if } i \in S, j \in R, z^i = z^j \tag{30.4}$$

$$\mu_i + \lambda_j\tilde{\alpha}_i + \frac{\tilde{\alpha}_i}{\alpha_j}\gamma_j{}'z^i \geq 0, \text{ if } i \in S, j \in R, z^i \neq z^j \tag{30.5}$$

5.3.2. Parametric representation of stationary points. a) For given, fixed parameters y, μ_i, λ_j, γ_j and ρ, (30) represents a system of linear equalities/inequalities for x. If rank $\mathbb{A}=n$, then (30) is reduced to (30.2)-(30.5) with $y=x$. b) The elimination of μ_i, $i \in S$, in (30.4) and (30.5) yields the equivalent conditions

$$(\lambda_j - \lambda_i) + \left(\frac{\gamma_j}{\alpha_j} - \frac{\gamma_i}{\alpha_i}\right)'z^i = 2\left(\frac{1}{\alpha_i} - \frac{1}{\alpha_j}\right), \text{ if } i,j \in R, z^i = z^j \tag{30.4a}$$

$$(\lambda_j - \lambda_i) + \left(\frac{\gamma_j}{\alpha_j} - \frac{\gamma_i}{\alpha_i}\right)'z^i \geq \frac{2}{\alpha_i}, \text{ if } i,j \in R, z^i \neq z^j \tag{30.5a}$$

Replacing (30.4),(30.5) by the equivalent conditions (30.4a), (30.5a), by means of Theorem 5.1 we obtain a D-stationarity characterization not involving S. Adding the inequalities (30.5a) corresponding to the index pairs (i,j), (j,i) with $z^i \neq z^j$, we obtain

$$\left(\frac{\gamma_j}{\alpha_j} - \frac{\gamma_i}{\alpha_i}\right)'(z^i - z^j) \geq 2\left(\frac{1}{\alpha_i} + \frac{1}{\alpha_j}\right), \text{ if } i,j \in R, z^i \neq z^j. \tag{31}$$

Therefore, we must have

i) if $z^i \neq z^j$, then $\gamma_j \neq \frac{\alpha_j}{\alpha_i}\gamma_i$,

ii) if $\gamma_j = \frac{\alpha_j}{\alpha_i}\gamma_i$, then $z^i = z^j$;

note that $z^i - z^j = (A^i - A^j)x - (b^i - b^j)$. c) If $\rho=0$, which holds e.g. if $D=\mathbb{R}^n$ or if y,x, resp., lies in the topological interior $\overset{o}{D}$ of D, then (30.2) is reduced to

$$\mathbb{A}'\gamma = \sum_{j \in R} A^{j\,\prime}\gamma_j = 0, \tag{32}$$

where $\gamma' = (\gamma_1', \gamma_2', \ldots, \gamma_r') \in \mathbb{R}^{r\cdot m}$. d) If the intersection of the linear subspaces of $\mathbb{R}^n$, generated by the row vectors $\{A_i^j: j \in R, 1 \leq i \leq m\}$ and $\{G_k: 1 \leq k \leq \kappa\}$, is the null space $\{0\}$, then (30.2) can only hold in the form $\mathbb{A}'\gamma = 0 = G'\rho$. Hence, also in this case γ must satisfy (32). e) Suppose that γ must fulfill the system of linear equations (32). Consequently, γ may be described by

$$\gamma_j = H^j c, \; c \in \mathbb{R}^q, \; q = r\cdot m - \operatorname{rank} \mathbb{A}, j \in R, \tag{33}$$

where $\operatorname{rank} \mathbb{A} \leq \min\{n, rm\}$ and $H^1, \ldots, H^r$ are fixed mxq matrices such that $H = \begin{pmatrix} H^1 \\ \vdots \\ H^r \end{pmatrix}$ has rank H=q. Putting (33) into (30.4a),(30.5a) and (31), we obtain

$$(\lambda_j - \lambda_i) + c'\left(\frac{H^j}{\alpha_j} - \frac{H^i}{\alpha_i}\right)' z^i = 2\left(\frac{1}{\alpha_i} - \frac{1}{\alpha_j}\right), \text{ if } i,j \in R, z^i = z^j,$$

$$(\lambda_j - \lambda_i) + c'\left(\frac{H^j}{\alpha_j} - \frac{H^i}{\alpha_i}\right)' z^i \geq \frac{2}{\alpha_i}, \text{ if } i,j \in R, \; z^i \neq z^j$$

as well as

$$c'\left(\frac{H^j}{\alpha_j} - \frac{H^i}{\alpha_i}\right)'(z^i - z^j) \geq 2\left(\frac{1}{\alpha_i} + \frac{1}{\alpha_j}\right), \text{ if } i,j \in R, z^i \neq z^j,$$

where c is now an arbitrary q-vector. In the special case q=0, i.e. if

$$r\cdot m = \operatorname{rank} \mathbb{A} (\leq n), \tag{34}$$

then c=0, and $\gamma = 0$ is the unique solution of (32). Because of (31), in case (34) we find that

$$z^1 = z^2 = \ldots = z^r. \tag{35}$$

Hence, if (34) holds, then a D-stationary point x lies in the linear manifold defined by (35). Note that in the case q=0 the system of linear equations $\mathbb{A}y = w$ has always a solution $y \in \mathbb{R}^n$ for every $w \in \mathbb{R}^{r\cdot m}$. f) In many practical cases only a few elements of the random matrix $(A(\omega), b(\omega))$ are stochastic. Hence, after a certain reordering of its elements, we then may get the representation

$$(A(\omega),b(\omega)) = (A_I(\omega),B,b(\omega)) \tag{36.1}$$

or

$$(A(\omega),b(\omega)) = \begin{pmatrix} A_I(\omega),b_I(\omega) \\ B \end{pmatrix}, \tag{36.2}$$

where $A_{II}=B$ resp. $(A_{II},b_{II})=B$ is a fixed $m \times \nu$ resp. $\mu \times (n+1)$ matrix. It is then easy to see that the stationarity of a point x is determined mainly by the random part of $(A(\omega),b(\omega))$.

5.3.3. Conditions for stationary points by means of $(\tilde{P}^N_{x,D})$. In addition to the characterization of D-stationary points be means of $(\tilde{P}^Q_{x,D})$, see Theorem 5.1, a similar stationarity condition, which is at least necessary, can be derived by means of $(\tilde{P}^N_{x,D})$.

Theorem 5.2. Let D be given by (23) and (27). If $x \in D$ is D-stationary, then there are an n-vector y, numbers μ_i, λ_j, ρ_k, $i \in S$, $j \in R$, $1 \le k \le \kappa$, and m-vectors γ_j, $j \in R$, such that

$$A^j y = A^j x \text{ for every } j \in R \text{ (resp. } y=x, \text{ if rank } \mathbb{A}=n) \tag{37.1}$$

$$0 = -\mathbb{A}'\gamma + G'\rho \tag{37.2}$$

$$Gx \le g \tag{37.3a}$$

$$Gy \le g,\ \rho'(Gy-g)=0,\ \rho \ge 0 \tag{37.3b}$$

$$\mu_i + \lambda_j \tilde{\alpha}_i + \frac{\tilde{\alpha}_i}{\alpha_j} \gamma_j' z^i = 0, \text{ if } i \in S, j \in R, z^i=z^j \tag{37.4}$$

$$\mu_i + \lambda_j \tilde{\alpha}_i + \frac{\tilde{\alpha}_i}{\alpha_j} \gamma_j' z^i \ge 0, \text{ if } i \in S, j \in R, z^i \ne z^j. \tag{37.5}$$

Proof. Let $N_o = \sum_{j \in R} A^{j\prime} A^j$. According to Theorem 4.3.1 x is D-stationary if and only if $(\tilde{P}^N_{x,D})$ has an optimal solution (y^*,T^*) such that $N(y^*) = \sum_{j \in R} ||A^j y^* - A^j x||_E^2 = (y^*-x)'N_o(y^*-x) = 0$ and $T^*=T^o$. From [7], page 108, we then know that there are still Lagrange multipliers μ_i^*, λ_j^*, ρ_k^*, γ_j^* such that the optimality conditions (29) hold with

$\tilde{C}=-N$. Since $N(y^*)=0$ implies that $N_o(y^*-x)=0$, the necessary conditions (37) for a D-stationary point x follow from (29) and (24),(25).

Eliminating $\mu_i, i \in S$, in (37), corresponding to the considerations in section 5.3.2, we find that (37.4),(37.5) may be replaced by the equivalent conditions

$$(\lambda_j-\lambda_i) + (\frac{\gamma_j}{\alpha_j} - \frac{\gamma_i}{\alpha_i})'z^i = 0, \text{ if } i,j \in R,\ z^i=z^j \quad (37.4a)$$

$$(\lambda_j-\lambda_i) + (\frac{\gamma_j}{\alpha_j} - \frac{\gamma_i}{\alpha_i})'z^i \geq 0, \text{ if } i,j \in R,\ z^i \neq z^j. \quad (37.5a)$$

From (37.5a) we then obtain

$$(\frac{\gamma_j}{\alpha_j} - \frac{\gamma_i}{\alpha_i})'(z^i-z^j) \geq 0, \text{ if } i,j \in R,\ z^i \neq z^j. \quad (38)$$

Comparing (30) and (37), we observe that (30.4a) and (30.5a) can be written in the form

$$(\lambda_j + \frac{2}{\alpha_j}) - (\lambda_i + \frac{2}{\alpha_i}) + (\frac{\gamma_j}{\alpha_j} - \frac{\gamma_i}{\alpha_i})'z^i = 0, \text{ if } z^i=z^j \quad (30.4b)$$

$$(\lambda_j + \frac{2}{\alpha_j}) - (\lambda_i + \frac{2}{\alpha_i}) + (\frac{\gamma_j}{\alpha_j} - \frac{\gamma_i}{\alpha_i})'z^i \geq \frac{2}{\alpha_j}, \text{ if } z^i \neq z^j. \quad (30.5b)$$

It is now easy to see that (30) implies (37). Note that the above conditions (37),(38) may be discussed in the same way as the conditions (30),(31) in section 5.3.2.

6. Optimal solutions (y^*,T^*) of $(\tilde{P}^Q_{x,D})$ having $\tau^*_{ij}>0$ for all $i\in S, j\in R$

Suppose again that $R=\{1,...,r\}$ is finite. Contained in the 4th part of the basic systems of relations (3),(5),(9),(10),(12) is always a condition of the type

$$1>\tau_{ij}>0 \text{ for at least one pair } (i,j), i\in S, j\in R \tag{39a}$$

or

$$\tau_{ij}>0 \text{ for all } i\in S \text{ and some } j\in R. \tag{39b}$$

Therefore, in this section we consider first the form of the matrix $T^* = (\tau_{ij})$ in an optimal solution (y^*,T^*) of $(\tilde{P}^Q_{x,D})$ fulfilling the stronger condition

$$\tau_{ij}>0 \text{ for all } i\in S, j\in R. \tag{39c}$$

Supposing (39c), from (29.2) and (25),(26) for τ_{ij} we obtain

$$0 = \frac{\partial\tilde{L}}{\partial\tau_{ij}} = \frac{\tilde{\alpha}_i}{\alpha_j}\, 2\, \frac{\tilde{\alpha}_i\,\tau_{ij}}{\alpha_j} + \mu_i + \lambda_j\,\tilde{\alpha}_i + \frac{\tilde{\alpha}_i}{\alpha_j}\,\gamma'_j z^i,$$

hence

$$\tau_{ij} = -\frac{1}{2}\left(\frac{\alpha_j}{\tilde{\alpha}_i}\right)^2 \mu_i - \frac{1}{2}\,\frac{\alpha_j^2}{\tilde{\alpha}_i}\,\lambda_j - \frac{1}{2}\,\frac{\alpha_j}{\tilde{\alpha}_i}\,\gamma'_j\, z^i \tag{40}$$

for all $i\in S, j\in R$. Since (y^*,T^*) must also fulfill (29.4), i.e. (12.1)-(12.3), we then get

$$1 = \sum_{j\in R}\tau_{ij} = -\frac{\mu_i}{2\tilde{\alpha}_i^2}\sum_{j\in R}\alpha_j^2 - \frac{1}{2\tilde{\alpha}_i}\sum_{j\in R}\alpha_j^2\,\lambda_j - \frac{1}{2\tilde{\alpha}_i}\Big(\sum_{j\in R}\alpha_j\gamma_j\Big)'z^i$$

and therefore

$$\mu_i = -\frac{2\,\tilde{\alpha}_i^2}{\sum\limits_{j\in R}\alpha_j^2} - \tilde{\alpha}_i\,\frac{\sum\limits_{j\in R}\alpha_j^2\,\lambda_j}{\sum\limits_{j\in R}\alpha_j^2} - \tilde{\alpha}_i\,\frac{\Big(\sum\limits_{j\in R}\alpha_j\,\gamma_j\Big)'z^i}{\sum\limits_{j\in R}\alpha_j^2}. \tag{41}$$

Inserting (41) into (40), we obtain

$$\tau_{ij} = \frac{\alpha_j^2}{\sum_{j\in R}\alpha_j^2} + \frac{\alpha_j^2}{2\tilde{\alpha}_i}\cdot\frac{\sum_{j\in R}\alpha_j^2\lambda_j}{\sum_{j\in R}\alpha_j^2} + \frac{\alpha_j^2}{2\tilde{\alpha}_i}\cdot\frac{(\sum_{j\in R}\alpha_j\gamma_j)'z^i}{\sum_{j\in R}\alpha_j^2} \tag{42}$$

$$-\frac{1}{2}\cdot\frac{\alpha_j^2}{2\tilde{\alpha}_i}\lambda_j - \frac{1}{2}\frac{\alpha_j}{\tilde{\alpha}_i}\gamma_j{}'z^i.$$

Moreover, for every $j \in R$ we must have

$$\alpha_j = \sum_{i\in S}\tilde{\alpha}_i\tau_{ij} = \frac{\alpha_j^2}{\sum_{j\in R}\alpha_j^2} + \frac{s\,\alpha_j^2}{2}\frac{\sum_{j\in R}\alpha_j^2\lambda_j}{\sum_{j\in R}\alpha_j^2} + \frac{\alpha_j^2}{2}\frac{(\sum_{j\in R}\alpha_j\gamma_j)'\sum_{i\in S}z^i}{\sum_{j\in R}\alpha_j^2} \tag{43}$$

$$-\frac{s}{2}\alpha_j^2\lambda_j - \frac{\alpha_j}{2}\gamma_j{}'\sum_{i\in S}z^i,$$

where $s=|S|$ is the cardinality of S. Equations (42) and (43) yield

$$\tau_{ij} - \frac{\alpha_j}{s\,\tilde{\alpha}_i} = \left(1-\frac{1}{s\,\tilde{\alpha}_i}\right)\frac{\alpha_j^2}{\sum_{j\in R}\alpha_j^2} - \frac{1}{2}\left(\alpha_j\gamma_j-\alpha_j^2\frac{\sum_{j\in R}\alpha_j\gamma_j}{\sum_{j\in R}\alpha_j^2}\right)'\frac{1}{\tilde{\alpha}_i}\left(z^i-\frac{1}{s}\sum_{i\in S}z^i\right),$$

hence, we have shown these formulas:

Theorem 6.1. Suppose that $R=\{1,\dots,r\}$ is finite. If (y^*,T^*), $T^*=(\tau_{ij})$, is an optimal solution of $(\tilde{P}^Q_{x,D})$ such that $\tau_{ij}>0$ for every $i\in S, j\in R$, then τ_{ij} is given by

$$\tau_{ij} = c^o_{ij} - \frac{1}{2}u_i{}'v_j,\quad i\in S, j\in R, \tag{44.1}$$

where

$$c^o_{ij} = \frac{\alpha_j}{s\,\tilde{\alpha}_i} + \left(1-\frac{1}{s\,\tilde{\alpha}_i}\right)\frac{\alpha_j^2}{\sum_{j\in R}\alpha_j^2}, \tag{44.2}$$

$$u_i = \frac{1}{\tilde{\alpha}_i}\left(z^i - \frac{1}{s}\sum_{i\in S}z^i\right), \tag{44.3}$$

$$v_j = \alpha_j \gamma_j - \alpha_j^2 \frac{\sum_{i\in R} \alpha_j \gamma_j}{\sum_{i\in R} \alpha_j^2} . \tag{44.4}$$

It is easy to see that

$$\sum_{j\in R} c^o_{ij} = 1, \quad \sum_{i\in S} \tilde{\alpha}_i c^o_{ij} = \alpha_j, \; i \in S, j \in R, \tag{45.1}$$

$$\sum_{i\in S} \tilde{\alpha}_i u_i = 0 \tag{45.2}$$

$$\sum_{j\in R} v_j = 0. \tag{45.3}$$

Sufficient conditions for $c^o_{ij}>0$, $i \in S, j \in R$, are contained in the next lemma:

Lemma 6.1. a) If $\alpha_j = \frac{1}{r}$ for every $j \in R$, then $c^o_{ij} = \frac{1}{r}$ for every $i \in S, j \in R$. b) If $\min_{j\in R} \alpha_j + s \min_{i\in S} \tilde{\alpha}_i \geq 1$, then $c^o_{ij} > 0$ for every $i \in S$, $j \in R$.

Proof. a) In this case it is $\frac{\alpha_j^2}{\sum_{j\in R} \alpha_j^2} = \frac{1}{r}$, which then yields the assertion. b) According to (44.2) we have only to consider the indices $1 \leq i \leq s$ such that $\tilde{\alpha}_i < \frac{1}{s}$. Because of $\frac{\alpha_j^2}{\sum_{j\in R} \alpha_j^2} < 1$, for $\tilde{\alpha}_i < \frac{1}{s}$ we find that

$$c^o_{ij} > \frac{\alpha_j}{s\,\tilde{\alpha}_i} + 1 - \frac{1}{s\,\tilde{\alpha}_i} = \frac{1}{s\,\tilde{\alpha}_i} (\alpha_j + s\tilde{\alpha}_i - 1)$$

$$\geq \frac{1}{s\,\tilde{\alpha}_i} (\min_{j\in R} \alpha_j + s \min_{i\in S} \tilde{\alpha}_i - 1),$$

which proves the assertion.

Note. Since $\tilde{\alpha}_i \geq \alpha_i, i \in S$, it is $\min_{i\in S} \tilde{\alpha}_i \geq \min_{i\in S} \alpha_i \geq \min_{j\in R} \alpha_j$. Hence, the condition in Lemma 6.1b holds e.g. if $\min_{j\in R} \alpha_j \geq \frac{1}{1+s}$.

It is easy to see that by the above considerations it is also possible to find the form of $T^{*} = (\tau_{ij})$ in an optimal solution (y^{*},T^{*}) of $(\tilde{P}^{Q}_{x,D})$ such that $\tau_{i_k j_k} = 1$ for given, fixed pairs $(i_k,j_k) \in S\times R$, $k=1,\ldots,k_0$.

Since each solution (y,T) of (12.1)-(12.4a),(12.4b) satisfies $\tau_{ij}>0$ for some pairs $(i,j) \in S\times R$, in section 8 we will apply formula (44) in the construction of solutions (y,T) of (12.1)-(12.4).

7. Existence of solutions of the SD-conditions (3.1)-(3.5), (12.1)-(12.5), resp.; Representation of stationary points

7.1. Solutions of (3.1)-(3.3),(3.5), (12.1)-(12.3),(12.5), resp.

7.1.1. Matrix representation (46.1)-(46.3) of (3.1)-(3.3) and (47.1)-(47.3) of (12.1)-(12.3). For simplification we again suppose that $R = \{1,2,\dots,r\}$ is finite. Define $1_r = (1,1,\dots,1)' \in \mathbb{R}^r$, $\alpha = (\alpha_1,\alpha_2,\dots,\alpha_r)'$ and let $B = (\beta_{ij})$ denote an $r\times r$ matrix. Considering especially $\beta_{ij} = \frac{\alpha_i \pi_{ij}}{\alpha_j}$, $i,j \in R$, one observes that the first three conditions of (3) can also be represented in the form

$$1_r'B = 1_r', \quad B \geq 0 \tag{46.1}$$

$$B\alpha = \alpha \tag{46.2}$$

$$Z_y = Z_x B, \tag{46.3}$$

where Z_y is the $m\times r$ matrix

$$Z_y = (A^1 y-b^1, A^2 y-b^2, \dots, A^r y-b^r), \quad y \in \mathbb{R}^n.$$

Similarly, considering $\tilde{\beta}_{ij} = \frac{\tilde{\alpha}_i \tau_{ij}}{\alpha_j}$, $i \in S$, $j \in R$, the corresponding conditions (12.1)-(12.3) can be given in the form

$$1_s'\tilde{B} = 1_r', \quad \tilde{B} \geq 0 \tag{47.1}$$

$$\tilde{B}\alpha = \tilde{\alpha} \tag{47.2}$$

$$Z_y = (A^{i_1}x-b^{i_1}, \dots, A^{i_s}x-b^{i_s})\tilde{B}, \tag{47.3}$$

where $1_s = (1,1,\dots,1)' \in \mathbb{R}^s$, $\tilde{\alpha} = (\tilde{\alpha}_{i_1}, \tilde{\alpha}_{i_2}, \dots, \tilde{\alpha}_{i_s})'$, $S = S_x = \{i_1, i_2, \dots, i_s\}$ and $\tilde{B} = (\tilde{\beta}_{ij})_{i \in S, j \in R}$ is an $s\times r$ matrix.

If $(\hat{\mathbb{A}}, \hat{\mathbb{b}})$, $(\tilde{\mathbb{A}}, \tilde{\mathbb{b}})$, U_B, $\tilde{U}_{\tilde{B}}$ denote the matrices

$$(\hat{\mathbb{A}},\hat{\mathbb{b}}) = \begin{pmatrix} A_1^1 & b_1^1 \\ \vdots & \\ A_1^r & b_1^r \\ A_2^1 & b_2^1 \\ \vdots & \\ A_2^r & b_2^r \\ \vdots & \\ A_m^1 & b_m^1 \\ \vdots & \\ A_m^r & b_m^r \end{pmatrix}, \qquad (\tilde{\mathbb{A}},\tilde{\mathbb{b}}) = \begin{pmatrix} A_1^{i_1} & b_1^{i_1} \\ \vdots & \\ A_1^{i_s} & b_1^{i_s} \\ A_2^{i_1} & b_2^{i_1} \\ \vdots & \\ A_2^{i_s} & b_2^{i_s} \\ \vdots & \\ A_m^{i_1} & b_m^{i_1} \\ \vdots & \\ A^{i_s} & b_m^{i_s} \end{pmatrix},$$

$$U_B = \begin{pmatrix} B' & 0 & \cdots & 0 \\ 0 & B' & \cdots & 0 \\ \vdots & & \ddots & \vdots \\ 0 & 0 & \cdots & B' \end{pmatrix}, \qquad \tilde{U}_{\tilde{B}} = \begin{pmatrix} \tilde{B}' & 0 & \cdots & 0 \\ 0 & \tilde{B}' & \cdots & 0 \\ \vdots & & \ddots & \vdots \\ 0 & 0 & \cdots & \tilde{B}' \end{pmatrix}$$

of sizes $m\cdot r \times (n+1)$, $m\cdot s \times (n+1)$, $m\cdot r \times m\cdot r$, $m\cdot r \times m\cdot s$, resp., then (46.3), (47.3) can also be represented in the form

$$\hat{\mathbb{A}}y = \hat{\mathbb{b}} + U_B(\hat{\mathbb{A}}x - \hat{\mathbb{b}}), \tag{46.3a}$$

$$\hat{\mathbb{A}}y = \hat{\mathbb{b}} + \tilde{U}_{\tilde{B}}(\tilde{\mathbb{A}}x - \tilde{\mathbb{b}}), \tag{47.3a}$$

resp.; note that $\hat{\mathbb{A}}$ and $\mathbb{A}$ are related by a simple row permutation, cf. (18), furthermore, if $m=1$, then $U_B = B'$ and $\tilde{U}_{\tilde{B}} = \tilde{B}'$.

Now, we consider our systems (46),(47) in more detail.

7.1.2. <u>Conditions (46.1),(46.2).</u> It is easy to see that $1_r'B = 1_r'$, $B\alpha = \alpha$ is a system of $2r$ linear equations having rank $2r-1$ for the r^2 elements β_{ij} of B. Moreover, the solutions B of (46.1), (46.2) may be represented by $B = I + \Theta$, where Θ is an rxr matrix such that $1_r'\Theta = 0$, $\Theta\alpha = 0$, $\Theta \geq -I$.

Let now $\mathcal{B}_\alpha$ denote the set of solutions B of (46.1), (46.2), hence

$$\mathcal{B}_\alpha = \{B: 1_r'B = 1_r', B\alpha = \alpha, B \geq 0\}.$$

If $B, C \in \mathcal{B}_\alpha$, then BC and CB are also elements of $\mathcal{B}_\alpha$. Since $\mathcal{B}_\alpha$ is a compact convex polyhedron in the rxr matrix space $\mathbb{R}^{r\cdot r}$, the elements B of $\mathcal{B}_\alpha$ have the form

$$B = \sum_{\nu=1}^{N} \sigma_\nu B^{(\nu)}, \quad \sum_{\nu=1}^{N} \sigma_\nu = 1, \quad \sigma_\nu \geq 0, \quad \nu = 1,2,\ldots,N, \tag{48}$$

where $B^{(1)} = I$ (identity matrix), $B^{(2)},\ldots,B^{(N)}$ are the extreme points of $\mathcal{B}_\alpha$, which may be determined in principle by linear programming. Since (46.1), (46.2) has rank 2r-1, we know that $N \leq \binom{r^2}{2r-1}$; furthermore, the maximal, minimal number of zeros in each $B^{(\nu)}$ is $r(r-1),(r-1)^2$, respectively.

Using the one-to-one linear transformation $C = \Gamma B$ from the rxr matrix space $\mathbb{R}^{r\cdot r}$ onto itself defined by

$$c_{ij} = \beta_{ij}\, \alpha_j, \quad i,j=1,2,\ldots,r,$$

where $C = (c_{ij})$, one observes that there is a one-to-one affine relationship between $\mathcal{B}_\alpha$ and the convex polytope

$$\mathcal{C}_\alpha = \{C: 1'C = \alpha', C1_r = \alpha, C \geq 0\}$$

in $\mathbb{R}^{r\cdot r}$. Especially, $B^{(\nu)}$ is an extreme point of $\mathcal{B}_\alpha$ if and only if $C^{(\nu)} = \Gamma B^{(\nu)}$ is an extreme point of $\mathcal{C}_\alpha$. Moreover, the conditions defining $\mathcal{C}_\alpha$ are exactly the constraints of a Hitchcock transportation problem. Knowing the extreme points of $\mathcal{C}_\alpha$, cf. [3],[20], by means of the transformation Γ, we now obtain this next lemma:

<u>Lemma 7.1.</u> A matrix $B = (\beta_{ij}) \in \mathcal{B}_\alpha$ is an extreme point of $\mathcal{B}_\alpha$

if and only if the connected components of the bipartite graph $G(B)$ are trees. Here, $G(B)$ is the (nondirected) bipartite graph with vertices $v_1, v_2, \ldots, v_r, w_1, w_2, \ldots, w_r$, where there are no edges between v_{i_1} and v_{i_2} or between w_{j_1} and w_{j_2}, $1 \le i_1, i_2 \le r$, $1 \le j_1, j_2 \le r$, and there is an edge between v_i and w_j if and only if $\beta_{ij} > 0, 1 \le i, j \le r$.

Examples. a) In the important special case

$$\alpha_j = \frac{1}{r} \text{ for every } j \in R$$

$\mathcal{B}_\alpha$ is the set of doubly stochastic matrices and $N = r!$. The extreme points $B^{(1)} = I, B^{(2)}, \ldots, B^{(r!)}$ of $\mathcal{B}_\alpha$ are just the permutation matrices of $\{1,2,\ldots,r\}$. b) For $r = 2$, and arbitrary $\alpha_1, \alpha_2 > 0$, $\alpha_1 + \alpha_2 = 1$, $\mathcal{B}_\alpha$ has the two extreme points

$$B^{(1)} = \begin{pmatrix} 1 & 0 \\ 0 & 1 \end{pmatrix}, \quad B^{(2)} = \begin{pmatrix} 1-t & t\frac{\alpha_1}{\alpha_2} \\ t & 1-t\frac{\alpha_1}{\alpha_2} \end{pmatrix}, \quad t = \min\{1, \frac{\alpha_2}{\alpha_1}\}.$$

c) Consequently, for arbitrary r and α, each matrix of the type

$$\begin{pmatrix}
1 & \cdots & 0 & \cdots & \cdots & 0 & \cdots & 0 \\
\vdots & \ddots & \vdots & & & \vdots & & \vdots \\
\vdots & 1 & \vdots & & & \vdots & & \vdots \\
0 & \cdots & 1-t & \cdots & \cdots & t\frac{\alpha_i}{\alpha_j} & \cdots & 0 \\
\vdots & & \vdots & \ddots & & \vdots & & \vdots \\
\vdots & & \vdots & 1 & & \vdots & & \vdots \\
\vdots & & \vdots & & \ddots & \vdots & & \vdots \\
0 & \cdots & t & \cdots & \cdots & 1-t\frac{\alpha_i}{\alpha_j} & \cdots & 0 \\
\vdots & & \vdots & & & \vdots & 1 & \vdots \\
\vdots & & \vdots & & & \vdots & \ddots & \vdots \\
0 & \cdots & 0 & \cdots & \cdots & 0 & \cdots & 1
\end{pmatrix}
\begin{matrix} \\ \\ \\ \ldots \text{ i-th row} \\ \\ \\ \\ \ldots \text{ j-th row} \\ \\ \\ \\ \end{matrix}
\tag{49.1}$$

(i-th column at the $1-t$ entry; j-th column at the $t\frac{\alpha_i}{\alpha_j}$ entry)

with

$$t = \min\{1, \frac{\alpha_j}{\alpha_i}\} \tag{49.2}$$

is an extreme point of $\mathcal{B}_\alpha$ for every choice of integers $1 \leq i,j \leq r$.

The related conditions (47.1), (47.2) may obviously be discussed in the same way: This system has rank $r+s-1$ and using here the one-to-one linear transformation $\tilde{C} = \tilde{\Gamma}\tilde{B}$ defined by $\tilde{c}_{ij} = \tilde{\beta}_{ij}\,\alpha_j$, $i \in S$, $j \in R$, where $\tilde{C} = (\tilde{c}_{ij})$, $\tilde{B} = (\tilde{\beta}_{ij})$, one obtains again the same representation for the extreme points $\tilde{B}^{(1)}, \tilde{B}^{(2)}, \ldots, \tilde{B}^{(\tilde{N})}$ of the convex polytope

$$\tilde{\mathcal{B}}_\alpha = \{\tilde{B}:\ 1_s' \tilde{B} = 1_r',\ \tilde{B}\alpha = \tilde{\alpha},\ \tilde{B} \geq 0\}$$

in $\mathbb{R}^{s \cdot r}$ as in Lemma 7.1. While $B^{(1)} = I$ is an extreme point of $\mathcal{B}_\alpha$, here we find, using formula (14), this lemma:

Lemma 7.2. For $i \in S, j \in R$ let $\tilde{\beta}_{ij}^{(1)} = 1$ if $z^i = z^j$ and $\tilde{\beta}_{ij}^{(1)} = 0$ if $z^i \neq z^j$. Then the matrix $\tilde{B}^{(1)} = (\tilde{\beta}_{ij}^{(1)})$ is an extreme point of $\tilde{\mathcal{B}}_\alpha$.

Proof. Obviously, $\tilde{B}^{(1)}$ is an element of $\tilde{\mathcal{B}}_\alpha$. Assume now that $\tilde{B}^{(1)} = \eta U + (1-\eta)V$ for elements $U = (u_{ij})$, $V = (v_{ij})$ of $\tilde{\mathcal{B}}_\alpha$ and $0<\eta<1$. If $i \in S, j \in R$ and $z^i \neq z^j$, then $0 = \tilde{\beta}_{ij}^{(1)} = \eta u_{ij} + (1-\eta)v_{ij}$, hence $u_{ij} = v_{ij} = 0$. Furthermore, if $i = q(j)$ denotes the unique integer in S such that $z^{q(j)} = z^j$, $j \in R$, then $1 = \tilde{\beta}_{ij}^{(1)} = \eta\, u_{ij} + (1-\eta)v_{ij}$, and therefore $u_{ij} = v_{ij} = 1$. Consequently, we find that $U = V = \tilde{B}^{(1)}$, which proves the assertion.

7.1.3. Condition (46.3). Since $A^j y - b^j = \sum_{i=1}^{r} (A^i x - b^i)\beta_{ij}$, where $\sum_{i=1}^{r} \beta_{ij} = 1$, $\beta_{ij} \geq 0$, $1 \leq i,j \leq r$, each m-vector $A^j y - b^j$ may be interpreted as a certain average of the vectors $A^1x-b^1, \ldots, A^r x-b^r$. If $\beta_1, \beta_2, \ldots, \beta_r$ denote the columns of the $r \times r$ matrix B, then the system of linear equations (46.3) for $(y, \beta_1, \ldots, \beta_r)$ may be given in the form

$$\begin{pmatrix} A^1 & -Z_x & & & \\ A^2 & & -Z_x & & \\ \vdots & & & \ddots & \\ A^r & & & & -Z_x \end{pmatrix} \begin{pmatrix} y \\ \beta_1 \\ \beta_2 \\ \vdots \\ \beta_r \end{pmatrix} = \begin{pmatrix} b^1 \\ b^2 \\ \vdots \\ b^r \end{pmatrix}, \qquad (50.1)$$

where Z_x is defined, see (46), by

$$Z_x = (A^1x-b^1, A^2x-b^2, \dots, A^rx-b^r). \qquad (50.2)$$

Comparing (50) with the constraints arising in stochastic linear programming with recourse, cf. [27],[63], we obtain:

Theorem 7.1. The linear system of equations (46.3) for (y,B) can be represented by (50). It has the dual decomposition matrix structure of a stochastic linear program with recourse having a discrete parameter distribution, where $W=-Z_x$ plays the role of the recourse matrix W.

Note. If rank $Z_x = m$, and there exists an r-vector e with positive components $e_1>o,\dots,e_r>o$ and $Z_xe=0$, then by [37] we know that $Z_x\beta = z$ has a solution $\beta\geq 0$ for every $z\in\mathbb{R}^m$.

Having this important representation (50) of (46.3), which opens very interesting algorithmic possibilities, cf. [27], we now study the solvability of (46.3),(47.3), which are equivalent to

(46.3a) $\hat{\mathbb{A}}y = \hat{\mathbb{b}} + U_B(\hat{\mathbb{A}}x - \hat{\mathbb{b}})$,

(47.3a) $\hat{\mathbb{A}}y = \hat{\mathbb{b}} + \tilde{U}_{\tilde{B}}(\tilde{\mathbb{A}}x - \tilde{\mathbb{b}})$, respectively.

I) For given $x\in\mathbb{R}^n$ this is a system of $m\cdot r$ linear equations for (y,B), $(y,\tilde{B})$, resp., containing $n + r^2$, $n+s\cdot r$ unknowns, respectively. Obviously, $(y,B) = (x,I)$, $(y,\tilde{B}) = (x,\tilde{B}^{(1)})$, resp., is always a solution of this system, where I is the rxr identity matrix and $\tilde{B}^{(1)}$ is defined in Lemma 7.2.

Theorem 7.2.1. a) For given $x \in \mathbb{R}^n$, the general solution $\{(y,B)\}$, $\{(y,\tilde{B})\}$, resp., of the homogeneous linear system associated with (46.3a), (47.3a), resp., has dimension

$\dim\{(y,B)\} \geq n+r^2-m\cdot r$, $\quad \dim\{(y,\tilde{B})\} \geq n+s\cdot r-m\cdot r$, respectively.

b) If rank $\hat{\mathbb{A}}$(= rank $\mathbb{A}$) = $m\cdot r$, then (46.3), (47.3) has a solution $y \in \mathbb{R}^n$ for every given tuple (x,B), $(x,\tilde{B})$, respectively.

Proof. The first part is an immediate consequence from the theory of matrix equations. From the assumptions in (b) we obtain

$m\cdot r = \text{rank } \hat{\mathbb{A}} \leq \text{rank } (\hat{\mathbb{A}}, \hat{\mathbb{b}}+U_B(\hat{\mathbb{A}}x-\hat{\mathbb{b}})) \leq m\cdot r$,

which proves the assertion for (46.3).

Note. Since rank $\hat{\mathbb{A}} \leq n$, the rank condition rank $\hat{\mathbb{A}} = m\cdot r$ can hold only as long as $m\cdot r \leq n$.

II) Solutions y of (46.3a), (47.3a), resp., of the form $y=Cx$, $y = \tilde{C}x$ may be found in a very simple way if $(\mathbb{A},\mathbb{b})$ fulfills the following condition

$$U_B\hat{\mathbb{A}} = \hat{\mathbb{A}}C, \quad \tilde{U}_{\tilde{B}}\tilde{\mathbb{A}} = \hat{\mathbb{A}}\tilde{C}, \text{ resp.,} \tag{51.1}$$

$$U_B\hat{\mathbb{b}} = \hat{\mathbb{b}}, \quad \tilde{U}_{\tilde{B}}\tilde{\mathbb{b}} = \hat{\mathbb{b}}, \text{ resp.,} \tag{51.2}$$

for some fixed matrix $B \in \mathcal{B}_\alpha$, $\tilde{B} \in \tilde{\mathcal{B}}_\alpha$, resp., and a certain nxn matrix $C,\tilde{C}$, respectively. For given tuple $(B,C),(\tilde{B},\tilde{C})$, resp., (51.1) is a homogeneous system of rmn linear equations for the $r\cdot m\cdot n$ elements of $\hat{\mathbb{A}}$. If $\dim\{\mathbb{A}\}$ denotes the dimension of the general solution $\{\mathbb{A}\}$ of this system, then

$\dim\{\mathbb{A}\} \geq r\cdot m\cdot n - \text{rank}(51.1)$.

We observe that (51.1) and (46.2), (47.2), resp., imply that

$\bar{A} = \bar{A}C$, $\bar{A} = \bar{A}\tilde{C}$, respectively, where $\bar{A} = EA(\omega)$.

The first part of (51.2) means simply that $\hat{\mathbb{b}}$ must be an eigenvector of U_B with the eigenvalue one. In the special case $m = 1$ (51) is reduced to

$B'\mathbb{A} = \mathbb{A}C$, $\tilde{B}'\tilde{\mathbb{A}} = \mathbb{A}C$, resp., (51.1a)

$B'\mathbb{b} = \mathbb{b}$, $\tilde{B}'\tilde{\mathbb{b}} = \mathbb{b}$, respectively. (51.2a)

Under the above assumptions we now find this result:

Theorem 7.2.2. Suppose that $(\mathbb{A},\mathbb{b})$ fulfills (51) with a certain $B \in \mathcal{B}_\alpha$, $\tilde{B} \in \tilde{\mathcal{B}}_\alpha$ and an nxn matrix $C,\tilde{C}$, respectively. Then $y=Cx$, $y = \tilde{C}x$, resp., is a solution of (46.3a), (47.3a), resp., for every $x \in \mathbb{R}^n$.

Proof. From (51) follows that

$$\hat{\mathbb{b}} + U_B(\hat{\mathbb{A}}x-\hat{\mathbb{b}}) = \hat{\mathbb{b}} + U_B\hat{\mathbb{A}}x - U_B\hat{\mathbb{b}}$$
$$= \hat{\mathbb{b}} + \hat{\mathbb{A}}Cx - \hat{\mathbb{b}} = \hat{\mathbb{A}}y,$$

hence $y = Cx$ satisfies (46.3a) for every x. The assertion concerning (47.3a) follows in the same way.

Let $\mathbb{A}_{(t)}$, $t=1,\dots,p$, be a decomposition of $\mathbb{A}$ into m_tx n submatrices, hence $\sum_{t=1}^{p} m_t = m\cdot r$. Corresponding to this partition of $\mathbb{A}$ we have then a partition of U_B into m_tx $m\cdot r$ submatrices $U_B^{(t)}$; moreover, each $U_B^{(t)}$ may be decomposed further into m_tx m_τ submatrices $U_B^{(t,\tau)}$, $t,\tau=1,\dots,p$. Based on this decomposition of $\mathbb{A}$ and U_B, (51.1) can be represented by

$$\sum_{\tau=1}^{p} U_B^{(t,\tau)}\hat{\mathbb{A}}_{(\tau)} = \hat{\mathbb{A}}_{(t)}C,\ t=1,\dots,p. \qquad (51.1.1)$$

In the interesting special case $U^{(t,\tau)}=0$ for $t\neq\tau$ (51.1.1) takes the simple form

$$U_B^{(t,t)}\hat{\mathbb{A}}_{(t)} = \hat{\mathbb{A}}_{(t)}C,\ t=1,\dots,p.$$

III) Instead of considering (46.3a), for a given n-vector x and a given matrix $(\hat{\mathbb{A}},\hat{\mathbb{b}})$, as a linear system of equations for the tuple (y,B), conversely, (46.3a) can also be interpreted, for a given tuple (x,y,B) as a system of linear equations for $(\hat{\mathbb{A}},\hat{\mathbb{b}})$. In the same way, (47.3a) can be interpreted, for a given tuple $(x,y,\tilde{B})$, as a system of linear equations for $(\hat{\mathbb{A}},\hat{\mathbb{b}})$.

Moreover, if $(\hat{\mathbb{A}},\hat{\mathbb{b}})$ is decomposed into $(\tilde{\mathbb{A}},\tilde{\mathbb{b}})$ and a complementary submatrix $(\hat{\mathbb{A}}_{II},\hat{\mathbb{b}}_{II})$ of $(\hat{\mathbb{A}},\hat{\mathbb{b}})$, then, for a given tuple $(\tilde{\mathbb{A}},y,\tilde{B})$, (47.3a) is a linear system of equations for $(x,\hat{\mathbb{A}}_{II},\hat{\mathbb{b}})$. From this point of view we now achieve this theorem:

Theorem 7.2.3. a) For a given tuple (x,y,B), (46.3a) is a homogeneous system of $m\cdot r$ linear equations for the $r\cdot m(n+1)$ unknowns in $(\hat{\mathbb{A}},\hat{\mathbb{b}})$. If $L_{(x,y,B)}$ denotes its general solution, then $\dim L_{(x,y,B)} \geq m\cdot r\cdot n$. b) For given tuples (x^t,y^t,B^t), $t = 1,2,\ldots,p$, it is $\dim \bigcap_{t=1}^{p} L_{(x^t,y^t,B^t)} \geq m\cdot r(n+1-p)$. Furthermore, if $B^t=B$, $t=1,\ldots,p$, and $(\hat{\mathbb{A}},\hat{\mathbb{b}})\in \bigcap_{t=1}^{p} L_{(x^t,y^t,B^t)}$, then $(\hat{\mathbb{A}},\hat{\mathbb{b}})\in L_{(x,y,B)}$ for every $(x,y)\in \mathrm{conv}\{(x^t,y^t)\colon t=1,\ldots,p\}$; if $x^t=x$, $t=1,\ldots,p$, and $(\hat{\mathbb{A}},\hat{\mathbb{b}})\in \bigcap_{t=1}^{p} L_{(x,y^t,B^t)}$, then $(\hat{\mathbb{A}},\hat{\mathbb{b}})\in L_{(x,y,B)}$ for every $(y,B)\in \mathrm{conv}\{(y^t,B^t)\colon t=1,\ldots,p\}$.

Proof. Given (x,y,B), obviously, (46.3a) is a homogeneous linear system of $m\cdot r$ equations for the $m\cdot r(n+1)$ unknowns in $(\hat{\mathbb{A}},\hat{\mathbb{b}})$. Consequently, $\dim L_{(x,y,B)} \geq m\cdot r(n+1) - m\cdot r$. The rest of the assertion follows in the same way.

A similar result holds for (47.3a).

IV) Suppose that $\hat{\mathbb{A}}$ contains a regular nxn submatrix $\hat{\mathbb{A}}_I$ and let $\hat{\mathbb{A}}_{II}$ denote the complementary $(m\cdot r-n)$xn submatrix of $\hat{\mathbb{A}}$. As mentioned in (II), this decomposition of $\hat{\mathbb{A}}$ yields a decomposition of U_B into submatrices U_B^I, U_B^{II}, which are decomposed themselves into submatrices $U_B^{I,I}$, $U_B^{I,II}$, $U_B^{II,I}$, $U_B^{II,II}$. Finally, $\hat{\mathbb{b}}$ is partitioned into submatrices $\hat{\mathbb{b}}_I$, $\hat{\mathbb{b}}_{II}$. Now, (46.3a) can be written in the form

$$\begin{aligned} y &= \hat{\mathbb{A}}_I^{-1}(\hat{\mathbb{b}}_I+U_B^I(\hat{\mathbb{A}}x-\hat{\mathbb{b}})) \\ \hat{\mathbb{A}}_{II}y &= \hat{\mathbb{b}}_{II}+U_B^{II}(\hat{\mathbb{A}}x-\hat{\mathbb{b}}). \end{aligned} \qquad (46.3b)$$

Consequently, if rank $\hat{\mathbb{A}}$(= rank $\mathbb{A}$) = n, then (46.3) is solvable for $y \in \mathbb{R}^n$ if and only if x and B are related by

$$(\hat{\mathbb{A}}_{II}\hat{\mathbb{A}}_I^{-1}\hat{\mathbb{b}}_I - \hat{\mathbb{b}}_{II}) + (\hat{\mathbb{A}}_{II}\hat{\mathbb{A}}_I^{-1}U_B^I - U_B^{II})(\hat{\mathbb{A}}x - \hat{\mathbb{b}}) = 0. \qquad (52)$$

This is a system of m·r-n equations for (x,B) having always the trivial solution (x,B) = (x,I). The system

$$(\hat{\mathbb{A}}_{II}\hat{\mathbb{A}}_I^{-1}U_B^I - U_B^{II})\hat{\mathbb{A}}x = 0 \qquad (53)$$

associated with (52) is homogeneously linear with respect to each variable x,B.

Theorem 7.2.4a. For given B let L be the linear space of solutions $x \in \mathbb{R}^n$ of (53). If $n \le m \cdot r < 2n$, then $\dim L \ge 2n - m \cdot r$.

Proof. The assertion follows from $\dim L \ge n - (m \cdot r - n) = 2n - m \cdot r > 0$.

Using the decomposition of U_B^I, U_B^{II} into the submatrices $U_B^{I,I}$, $U_B^{I,II}$ and $U_B^{II,I}$, $U_B^{II,II}$, (52) can also be represented by

$$(\hat{\mathbb{A}}_{II}\hat{\mathbb{A}}_I^{-1}\hat{\mathbb{b}}_I - \hat{\mathbb{b}}_{II}) - (\hat{\mathbb{A}}_{II}\hat{\mathbb{A}}_I^{-1}U_B^{I,I} - U_B^{II,I})\hat{\mathbb{b}}_I - (\hat{\mathbb{A}}_{II}\hat{\mathbb{A}}_I^{-1}U_B^{I,II} - U_B^{II,II})\hat{\mathbb{b}}_{II} +$$

$$+(\hat{\mathbb{A}}_{II}\hat{\mathbb{A}}_I^{-1}U_B^{I,I}\hat{\mathbb{A}}_I - U_B^{II,I}\hat{\mathbb{A}}_I + \hat{\mathbb{A}}_{II}\hat{\mathbb{A}}_I^{-1}U_B^{I,II}\hat{\mathbb{A}}_{II} - U_B^{II,II}\hat{\mathbb{A}}_{II})x = 0. \qquad (52.a)$$

Several interesting simplifications of (52) may be obtained now by setting equal to zero some of the submatrices $U_B^{I,I}$, $U_B^{I,II}$, $U_B^{II,I}$, $U_B^{II,II}$ of U_B.

The next result is based on the following generalization of condition (51)

$$U_B^I\hat{\mathbb{A}} = \hat{\mathbb{A}}_I C_I \qquad (54.1)$$

$$U_B^{II}\hat{\mathbb{A}} = \hat{\mathbb{A}}_{II} C_{II} \qquad (54.2)$$

$$\hat{\mathbb{b}}_{II} = U_B^{II}\hat{\mathbb{b}} + \hat{\mathbb{A}}_{II} c_{II}, \qquad (54.3)$$

where C_I, C_{II} are certain nxn matrices and c_{II} is a certain n-vector.

Note. (54.1) implies that $C_I = \hat{A}_I^{-1} U_B^I \hat{A}$.

Theorem 7.2.4b. If (54.1) and (54.2) hold, then (53) has the form

$$\hat{A}_{II}(C_I - C_{II})x = 0. \tag{55.1}$$

If (54.1)-(54.3) is fulfilled, then (52) is implied by

$$(C_I - C_{II})x = -\hat{A}_I^{-1}\hat{b}_I + \hat{A}_I^{-1}U_B^I\hat{b} + c_{II} \tag{52.2}$$

and (55.2) is also necessary for (52) provided that rank $\hat{A}_{II} = n$.

Proof. a) The first two conditions in (54) yield

$$(\hat{A}_{II}\hat{A}_I^{-1}U_B^I - U_B^{II})\hat{A} = \hat{A}_{II}\hat{A}_I^{-1}\hat{A}_I C_I - \hat{A}_{II}C_{II} = \hat{A}_{II}(C_I - C_{II}).$$

b) If (54.1)-(54.3) hold, then the terms in (52), which are independent on x, have the form

$$(\hat{A}_{II}\hat{A}_I^{-1}\hat{b}_I - \hat{b}_{II}) - (\hat{A}_{II}\hat{A}_I^{-1}U_B^I - U_B^{II})\hat{b} =$$

$$= \hat{A}_{II}(\hat{A}_I^{-1}\hat{b}_I - \hat{A}^{-1}U_B^I\hat{b}) - \hat{A}_{II}c_{II} = \hat{A}_{II}(\hat{A}_I^{-1}\hat{b}_I - \hat{A}_I^{-1}U_B^I\hat{b} - c_{II}).$$

The assertion follows now from the equations shown above.

Note. An interesting situation occurs if $U_B^{I,II} = 0$. If this holds, then $\hat{A}_I^{-1}U_B^I\hat{b} = \hat{A}_I^{-1}U_B^{I,I}\hat{b}_I = \hat{A}_I^{-1}U_B^{I,I}\hat{A}_I\hat{A}_I^{-1}\hat{b}_I = \hat{A}_I^{-1}U_B^I\hat{A}\hat{A}_I^{-1}\hat{b}_I =$ $\hat{A}_I^{-1}\hat{A}_I C_I\hat{A}_I^{-1}\hat{b}_I = C_I\hat{A}_I^{-1}\hat{b}_I$.

V. For $x \in \mathbb{R}^n$ and $B \in \mathcal{B}_\alpha$ let $q^B(x)$ denote the left hand side of equation (52), i.e.

$$q^B(x) = (\hat{A}_{II}\hat{A}_I^{-1}\hat{b}_I - \hat{b}_{II}) + (\hat{A}_{II}\hat{A}_I^{-1}U_B^I - U_B^{II})(\hat{A}x - \hat{b}),$$

and let $q^{(\nu)}(x) = q^B(x)$ for $B = B^{(\nu)}$, $\nu = 1,\dots,N$, where $I = B^{(1)}$, $B^{(2)},\dots,B^{(N)}$ are the extreme points of $\mathcal{B}_\alpha$.

Theorem 7.2.5. Equation (52) has a solution $(x,B) \in \mathbb{R}^n \times \mathcal{B}_\alpha$ such that $B \neq I$, i.e. nontrivial solution, if and only if

$$q^{(\nu)}(x)'u < 0,\ \nu = 2,\dots,N \tag{56}$$

has no solution $u \in \mathbb{R}^{r\cdot m-n}$.

Proof. Since $B \to q^B(x)$ is affine-linear and $q^{(1)}(x) = 0$ for every $x \in \mathbb{R}^n$, (52) is equivalent to $\sum_{\nu=2}^{N} \sigma_\nu q^{(\nu)}(x) = 0$, where $\sigma_2,\ldots,\sigma_N$ are nonnegative coefficients such that $\sum_{\nu=2}^{N} \sigma_\nu \leq 1$. Since this inequality is redundant, the assertion now follows from the transposition theorem of Gordan, cf. [54].

A simple consequence is this corollary:

<u>Corollary 7.1.</u> If for $x \in \mathbb{R}^n$ there are integers $2 \leq \mu, \kappa \leq N$ such that $q^{(\mu)}(x) = - q^{(\kappa)}(x)$, then there exists $B \in \mathcal{B}_\alpha$, $B \neq I$, such that (x,B) is a solution of (52).

This corollary may be applied e.g. if $U_B \hat{\mathbb{b}} = \hat{\mathbb{b}}$ and $\mathbb{A}$ can be decomposed into nxn submatrices $\mathbb{A}_{(1)}, \mathbb{A}_{(2)}, \ldots, \mathbb{A}_{(p)}$, where $\mathbb{A}_{(1)} = \mathbb{A}_I$ and in the corresponding decomposition of U_B, cf. (51.1.1), we have that $U_B^{(t,\tau)} = 0$ for $t \neq \tau$. Then $q^B(x)$ takes the form

$$q^B(x) = \begin{pmatrix} (\hat{\mathbb{A}}_{(2)} \hat{\mathbb{A}}_{(1)}^{-1} & U_B^{(1,1)} \hat{\mathbb{A}}_{(1)} & - U_B^{(2,2)} \hat{\mathbb{A}}_{(2)})x \\ \cdot & \cdot & \cdot \\ \cdot & \cdot & \cdot \\ \cdot & \cdot & \cdot \\ (\hat{\mathbb{A}}_{(p)} \hat{\mathbb{A}}_{(1)}^{-1} & U_B^{(1,1)} \hat{\mathbb{A}}_{(1)} & - U_B^{(p,p)} \hat{\mathbb{A}}_{(p)})x \end{pmatrix}$$

If μ,κ are integers such that $2 \leq \mu,\kappa \leq N$, $\mu \neq \kappa$, and

$$U_{B(\nu)}^{(1,1)} \hat{\mathbb{A}}_{(t)} = \hat{\mathbb{A}}_{(t)} U_{B(\nu)}^{(1,1)}, \quad t=1,\ldots,p, \quad \nu = \mu,\kappa,$$

then we obtain

$$q^{(\nu)}(x) = \begin{pmatrix} (U_{B(\nu)}^{(1,1)} - U_{B(\nu)}^{(2,2)}) \hat{\mathbb{A}}_{(2)} x \\ \cdot \\ \cdot \\ \cdot \\ (U_{B(\nu)}^{(1,1)} - U_{B(\nu)}^{(p,p)}) \hat{\mathbb{A}}_{(p)} x \end{pmatrix} \quad \text{for } \nu = \mu,\kappa.$$

Consequently, if $U_{B(\nu)}^{(t,t)} = U_{B(\nu)}^{(2,2)}$, $t=3,\ldots,p$, $\nu = \mu,\kappa$ and $U_{B(\mu)}^{(1,1)} = U_{B(\kappa)}^{(2,2)}$, $U_{B(\kappa)}^{(1,1)} = U_{B(\mu)}^{(2,2)}$, then $q^{(\mu)}(x) = - q^{(\kappa)}(x)$.

7.1.4. Geometrical representation of (46), (47). Using (48), we find that the basic system (46) is equivalent to

$$Z_y = \sum_{\nu=1}^{N} \sigma_\nu \, Z_x B^{(\nu)}$$
$$\sum_{\nu=1}^{N} \sigma_\nu = 1, \; \sigma_\nu \geq 0, \; \nu = 1,\dots,N. \tag{57}$$

For given n-vector x, (57) is a condition for (y,σ), $\sigma=(\sigma_1,\dots,\sigma_N)'$. A corresponding representation holds for (47). Geometrically, (57) describes the intersection of the affine subspace

$$V = \{Z_y : y \in \mathbb{R}^n\}$$

of $\mathbb{R}^{m\cdot r}$ with the convex polytope

$$P_x = \mathrm{conv}\{Z_x, Z_x B^{(2)}, \dots, Z_x B^{(N)}\}$$

of Z_x - averages generated by the mxr matrices $Z_x B^{(\nu)}, \nu=1,\dots,N$. Hence, we have that

$$\{y \in D\text{: There is an rxr matrix } \Pi, B\text{, resp., such that } (y,\Pi), (y,B)\text{, resp., fulfills (3.1)-(3.3), (46), resp.}\} = \{y \in \mathbb{R}^n : Z_y \in V_D \cap P_x\}, \tag{58}$$

where $V_D = \{Z_y : y \in D\}$ is the embedding of D into V. Obviously, if D is convex, then V_D is also convex.

We observe that the convex polytope P_x lies on the linear manifold

$$\sum_{j=1}^{r} \alpha_j v^j = \sum_{i=1}^{r} \alpha_i (A^i x - b^i)$$

of $\mathbb{R}^{m\cdot r}$, where $v^1, v^2, \dots, v^r$ are the columns of the elements $(v^1, v^2, \dots, v^r)$ of P_x. Moreover, we have this lemma:

Lemma 7.4. Z_x is an extreme point of P_x.

Proof. Suppose that $Z_x = \eta C + (1-\eta)\Gamma$, where $0<\eta<1$ and $C, \Gamma \in P_x$, hence $C = \sum_{\nu=1}^{N} \lambda_\nu Z_x B^{(\nu)}$, $\Gamma = \sum_{\nu=1}^{N} \mu_\nu Z_x B^{(\nu)}$ with $\lambda_\nu \geq 0$, $\mu_\nu \geq 0$, $1 \leq \nu \leq N$, and $\sum_{\nu=1}^{N} \lambda_\nu = \sum_{\nu=1}^{N} \mu_\nu = 1$. Consequently,

$$Z_x = \eta \sum_{\nu=1}^{N} \lambda_\nu Z_x B^{(\nu)} + (1-\eta) \sum_{\nu=1}^{N} \mu_\nu Z_x B^{(\nu)} = Z_x B,$$

where $B = (b_{ij})$ is defined by

$$B = \sum_{\nu=1}^{N} (\eta\lambda_\nu + (1-\eta)\mu_\nu) B^{(\nu)};$$

moreover, it is $B \in \mathcal{B}_\alpha$ and

$$A^j x - b^j = \sum_{i \in R} (A^i x - b^i) b_{ij} = \sum_{i \in S} (A^i x - b^i) \tilde{b}_{ij},$$

where $\tilde{b}_{ij} = \sum_{z^t = z^i} b_{tj}$, $i \in S, j \in R$. Since $\sum_{i \in S} \tilde{b}_{ij} = 1$, $\sum_{j \in R} \tilde{b}_{ij} \alpha_j = \tilde{\alpha}_i$, $\tilde{b}_{ij} \geq 0$, $i \in S, j \in R$, according to the proof of Theorem 4.1 we know that

$$\sum_{z^t = z^i} b_{tj} = \tilde{b}_{ij} = \begin{cases} 0, & \text{if } z^i \neq z^j \\ 1, & \text{if } z^i = z^j \end{cases} \quad \text{for all } i \in S, j \in R.$$

Denoting the elements of $B^{(\nu)}$ by $b_{ij}^{(\nu)}$, the definition of B yields now

$$\eta \sum_{\nu=1}^{N} \lambda_\nu \sum_{z^t = z^i} b_{tj}^{(\nu)} + (1-\eta) \sum_{\nu=1}^{N} \mu_\nu \sum_{z^t = z^i} b_{tj}^{(\nu)} = \begin{cases} 0, & z^i \neq z^j \\ 1, & z^i = z^j \end{cases}, \quad i \in S, j \in R.$$

Because of $0<\eta<1$, $\lambda_\nu \geq 0$, $\mu_\nu \geq 0$ and $0 \leq \sum_{z^t = z^i} b_{tj}^{(\nu)} \leq 1$ we must therefore have

$$\sum_{\nu=1}^{N} \lambda_\nu \sum_{z^t = z^i} b_{tj}^{(\nu)} = \sum_{\nu=1}^{N} \mu_\nu \sum_{z^t = z^i} b_{tj}^{(\nu)} = \begin{cases} 0, & z^i \neq z^j \\ 1, & z^i = z^j \end{cases}, \quad i \in S, j \in R.$$

Since $C = \sum_{\nu=1}^{N} \lambda_\nu Z_x B^{(\nu)} = Z_x(\sum_{\nu=1}^{N} \lambda_\nu B^{(\nu)})$ and $\Gamma = Z_x(\sum_{\nu=1}^{N} \mu_\nu B^{(\nu)})$, for the jth column c_j, d_j of C, Γ, resp., we obtain

$$c_j = \sum_{i\in R} z^i\left(\sum_{\nu=1}^{N} \lambda_\nu\, b_{ij}^{(\nu)}\right) = \sum_{i\in S} z^i\left(\sum_{\nu=1}^{N} \lambda_\nu \sum_{z^t=z^i} b_{tj}^{(\nu)}\right) =$$

$$= z^{q(j)} = z^j$$

as also $d_j = z^j$, where $i=q(j)$ is the unique index in S such that $z^{q(j)} = z^j$. Hence, $C = \Gamma = Z_x$, which now proves that Z_x is indeed an extreme point of P_x.

7.1.5. Necessary conditions for (3.1)-(3.3), (46.1)-(46.3). If $x,y\in\mathbb{R}^n$ are related by means of one of the equivalent conditions (3.1)-(3.3), (46.1)-(46.3), resp., then $\bar{A}y=\bar{A}x$, see Lemma 2.1. Considering, besides the mean $\bar{A}x-\bar{b}$ of the random vector $A(\omega)x-b(\omega)$, also its covariance matrix

$$Q_x = \operatorname{var}(A(\cdot)x-b(\cdot)) = \sum_{j=1}^{r} \alpha_j (A^jx-b^j-(\bar{A}x-\bar{b}))(A^jx-b^j-(\bar{A}x-\bar{b}))',$$

we find this theorem:

Theorem 7.3. a) If $x,y\in\mathbb{R}^n$ are related by virtue of (3.1)-(3.3), (46.1)-(46.3), resp., then

$$\bar{A}x = \bar{A}y \tag{59.1}$$

$$Q_x - Q_y \text{ is positive semidefinite.} \tag{59.2}$$

b) If in addition (3.4a) holds, i.e. if the transition probability measure $K^j = \sum_{i\in R} \beta_{ij}\, \varepsilon_{z^i}$, $K^j = \sum_{i\in R} b_{ij}\, \varepsilon_{z^i}$, resp., is not degenerated for at least one $j\in R$, then (59) holds, where $Q_y \neq Q_x$.

c) If there exist additionally $j\in R$ and $i_o,\dots,i_m\in R$ such that $b_{ij}>o$, $i=i_o,\dots,i_m$ and rank $((A^{i_1}x-b^{i_1})-(A^{i_o}x-b^{i_o}),\dots,(A^{i_m}x-b^{i_m})-(A^{i_o}x-b^{i_o})) = m$, then (59) is fulfilled, where Q_x-Q_y is positive definite.

Proof. a) Using (46.1)-(46.3), for every $z\in\mathbb{R}^m$ it is

$$z'Q_yz = \sum_{j=1}^{r} \alpha_j((A^jy-b^j-(\bar{A}x-\bar{b}))'z)^2 = \sum_{j=1}^{r} \alpha_j\Big(\sum_{i=1}^{r} b_{ij}(A^ix-b^i-$$

$$(\bar{A}x-\bar{b}))'z)^2 \le \sum_{j=1}^{r} \alpha_j \sum_{i=1}^{r} b_{ij}((A^i x-b^i-(\bar{A}x-\bar{b}))'z)^2 =$$

$$= \sum_{i=1}^{r} \alpha_i((A^i x-b^i-(\bar{A}x-\bar{b}))'z)^2 = z'Q_x z,$$

hence, Q_x-Q_y is positive semidefinite. b) If K^j is not degenerated for at least one $j \in R$, then a similar argument shows that $z'Q_y z<z'Q_x z$ for at least one $z\neq 0$, and therefore $Q_y \neq Q_x$.

c) In this case, K^j is not degenerated, and for every $z\neq 0$ there exist indices i_k, i_κ, $0\le k,\kappa\le m$, such that $(A^{i_k}x-b^{i_k})'z \neq (A^{i_\kappa}x-b^{i_\kappa})'z$, hence $z'Q_y z < z'Q_x z$.

Note. a) In [44] we have shown that for normally distributed random matrices $(A(\omega),b(\omega))$ (59) already implies that $F(y) \le F(x)$. b) If $\text{vec}(A,b) = (A_1,b_1,A_2,b_2,\dots,A_m,b_m)'$, where (A_i,b_i) is the i-th row of (A,b) and Δ_x denotes the $m(n+1)\text{x } m$ matrix

$$\Delta_x = \begin{pmatrix} \hat{x} & 0 & \dots & 0 \\ 0 & \hat{x} & \dots & 0 \\ \cdot & \cdot & & \cdot \\ \cdot & \cdot & \cdot & \cdot \\ \cdot & \cdot & & \cdot\,\cdot \\ 0 & 0 & \dots & \hat{x} \end{pmatrix}, \quad \hat{x} = \begin{pmatrix} x \\ -1 \end{pmatrix}, \; x \in \mathbb{R}^n,$$

then $Q_x = \text{var}(A(\cdot)x-b(\cdot)) = \Delta_x' \text{ var vec } (A(\cdot),b(\cdot))\Delta_x$. In the special case m=1 it is $Q_x = \hat{x}' \operatorname{var} \begin{pmatrix} A(\cdot)' \\ b \end{pmatrix} \hat{x}$.

7.1.6. Representation (60) of the random matrix $(A(\omega), b(\omega))$.
Obviously, the $m\times(n+1)$ random matrix $(A(\omega), b(\omega))$ can always be represented by

$$(A(\omega), b(\omega)) = (A^{(0)}, b^{(0)}) + \sum_{t=1}^{L} \xi_t(\omega)(A^{(t)}, b^{(t)}), \tag{60}$$

where $(A^{(0)}, b^{(0)}), (A^{(1)}, b^{(1)}), \ldots, (A^{(L)}, b^{(L)})$ are L+1 given fixed $m\times(n+1)$ matrices and $\xi_1(\omega), \ldots, \xi(\omega)$ are discretely distributed real valued random variables with mean

$$E\xi_t(\omega) = 0, \ t = 1,2,\ldots,L. \tag{60.1}$$

Hence, if $\xi_t^1, \xi_t^2, \ldots, \xi_t^r$ are the possible realizations of $\xi_t(\omega)$, then

$$(A^i, b^i) = (A^{(0)}, b^{(0)}) + \sum_{t=1}^{L} \xi_t^i (A^{(t)}, b^{(t)}),$$

and α_i is defined here by $\alpha_i = P(\xi_1(\omega) = \xi_1^i, \ldots, \xi_L(\omega) = \xi_L^i)$ for all $i \in R$. Moreover, $0 = E\xi_t(\omega) = \sum_{i \in R} \xi_t^i \alpha_i$ and $(\bar{A}, \bar{b}) = E(A(\omega), b(\omega)) = (A^{(0)}, b^{(0)})$.

We observe that in numerous practical applications the number L of one-dimensional random variables $\xi_t(\omega)$ contained in $(A(\omega), b(\omega))$ will be small relative to the dimension $m(n+1)$ of $(A(\omega), b(\omega))$.

Having

$$z^i = A^i x - b^i = A^{(0)}x-b^{(0)} + \sum_{t=1}^{L} (A^{(t)}x-b^{(t)})\xi_t^i, \; i \in R,$$

and

$$w^j = A^j y-b^j = A^{(0)}y-b^{(0)} + \sum_{t=1}^{L} (A^{(t)}y-b^{(t)})\xi_t^j, \; j \in R,$$

equation (46.3) can be written in the form

$$A^{(0)}y-b^{(0)} + \sum_{t=1}^{L} (A^{(t)}y-b^{(t)})\xi_t^j = w^j = \sum_{i\in R} z^i b_{ij} =$$

$$= \sum_{i\in R} (A^{(0)}x-b^{(0)}+ \sum_{t=1}^{L} (A^{(t)}x-b^{(t)})\xi_t^i)b_{ij} =$$

$$= A^{(0)}x-b^{(0)} + \sum_{t=1}^{L} (A^{(t)}x-b^{(t)})(\sum_{i\in R} \xi_t^i b_{ij}), \; j \in R,$$

where $B=(b_{ij})$ is defined by (46.1) and (46.2).

Because of (60.1) these relations are - see also (4) - equivalent to the equations

$$A^{(0)}y = A^{(0)}x \quad (\text{i.e. } \bar{A}y = \bar{A}x), \tag{61.1}$$

$$\sum_{t=1}^{L} (A^{(t)}y-b^{(t)})\xi_t^j = \sum_{t=1}^{L} (A^{(t)}x-b^{(t)})(\sum_{i\in R} \xi_t^i b_{ij}), \; j \in R. \tag{61.2}$$

If Ξ denotes the fixed Lxr matrix of ξ_t-realizations

$$\Xi = \begin{pmatrix} \xi_1^1 & \dots & \xi_1^i & \dots & \xi_1^r \\ \vdots & & \vdots & & \vdots \\ \xi_t^1 & \dots & \xi_t^i & \dots & \xi_t^r \\ \vdots & & \vdots & & \vdots \\ \xi_L^1 & \dots & \xi_t^i & \dots & \xi_L^r \end{pmatrix},$$

then (61.2) can be written in the matrix form

$$(A^{(1)}y-b^{(1)},\dots,A^{(L)}y-b^{(L)})\Xi = (A^{(1)}x-b^{(1)},\dots,A^{(L)}x-b^{(L)})\Xi B, \tag{61.2a}$$

furthermore, (60.1) is equivalent to

$$\Xi\alpha = 0. \tag{60.1a}$$

Clearly, we may suppose that $\Xi \neq 0$.

a) We consider first the case L=1, hence $(A(\omega),b(\omega))$ depends on a single real valued random variable $\xi_1(\omega)$. Using (61.2a), it is easy to see that (46) holds here if and only if there exist a matrix B satisfying (46.1), (46.2) and a number $h \in \mathbb{R}$ such that

$$\Xi B = h\Xi, \tag{62.1}$$

$$A^{(1)}y-b^{(1)} = h(A^{(1)}x-b^{(1)}), \quad A^{(0)}y = A^{(0)}x, \tag{62.2}$$

hence, Ξ must be a left eigenvector of B, and h is a left eigenvalue of B. We observe that, in the present case, there is only an implicit relationship - given by (62.1) and (62.2) - between the unknowns y and B.

b) Trying also in the general case $L \geq 1$ to separate y, B from each other - as found in (62) for the L=1 case - we see that solutions (y,B) of (46) may be obtained by solving this system of linear relations

$$1_r'B = 1_r', \quad B \geq 0 \tag{63.1}$$

$$B\alpha = \alpha \tag{63.2}$$

$$\Xi B = H\Xi \tag{63.3}$$

$$A^{(0)}y = A^{(0)}x \tag{63.4}$$

$$(A^{(1)}y-b^{(1)},\dots,A^{(L)}y-b^{(L)}) = (A^{(1)}x-b^{(1)},\dots,A^{(L)}x-b^{(L)})H \tag{63.5}$$

for (y,B,H), where $H = (h_{t\tau})_{t,\tau=1,\dots,L}$ is an auxiliary LxL matrix.

While the basic equation (46.3) contains $m \cdot r$ linear equations for y, we observe that (63.5) is a system of $m \cdot L$ linear equations for y, hence, the number of equations for y in (63) is independent of the number of realizations of $\xi(\omega)$. Note that (63.3) can be interpreted as a generalized eigenvalue problem with the Lxr "left eigenmatrix" Ξ and the generalized "left eigenvalue" H. If $H = (h_{tt}\,\delta_{t\tau})$ is a diagonal matrix, then (63.3) represents the following L ordinary eigenvalue problems

$$\Xi_t B = h_{tt}\,\Xi_t, \quad t=1,\dots,L, \tag{63.3.1}$$

where Ξ_t denotes the t-th row of Ξ, and (63.5) has then this simple form (see (62))

$$A^{(t)}y-b^{(t)}=h_{tt}(A^{(t)}x-b^{(t)}),\ t=1,\dots,L. \tag{63.5.1}$$

Note that (63.1) and (63.3.1) imply that $|h_{tt}|\le\sqrt{r}$, $t=1,\dots,L$.

7.1.6.1. Examples. In the following we use the fact that under the weak assumption

$\operatorname{rank} \Xi = L$

the equations $A^j y=A^j x$ hold for all $j \in R$ if and only if $A^{(t)}y = A^{(t)}x$ for all $t=0,1,\dots,L$.

Example 1. Consider a vector $x \in D$ with $A^{(t)}x \neq b^{(t)}$ for at least one $1\le t\le L$ and suppose that there is a $y_o \in \mathbb{R}^n$ such that $A^{(0)}x = A^{(0)}y_o$ and $A^{(t)}y_o = b^{(t)}$ for all $t=1,\dots,L$. Define then the rxr matrix B_o by $B_o = \alpha 1'$. Because of (60.1a) it is $\Xi B_o = 0$. Thus, the tuple (y_o,B_o) satisfies (46.1), (46.2), (61.1), (61.2a) and therefore also (46). Since $A^{(t)}y_o = b^{(t)} \neq A^{(t)}x$ for at least one $1\le t\le L$, we find $y_o\neq x$ and, if $\operatorname{rank} \Xi=L$, $A^j y_o\neq A^j x$ for at least one $j \in R$. Note that in the special case $\bar{A} = A^{(0)} = 0$ the vector y_o is determined only by the equations $A^{(t)}y=b^{(t)}$, $t=1,\dots,L$. If $x \in \overset{o}{D}$ (= interior of D), then there is a number $0<\lambda<1$ such that $y = \lambda y_o + (1-\lambda)x \in D$. Since (46) is linear with respect to (y,B), it easy to see that $(y,B) = \lambda(y_o,B_o) + (1-\lambda)(x,I)$ also satisfies (46), where now $x \in D$ and also $A^j y \neq A^j x$ for at least one $j \in R$, provided that $\operatorname{rank} \Xi=L$.

Example 2: Symmetric solutions of (63.1)-(63.3).
Suppose, for simplifications, that $\alpha_j = \frac{1}{r}$, $j \in R$, and that $\xi_1(\omega),\dots,\xi_L(\omega)$ are uncorrelated random variables. If B is a symmetric matrix and $H = (h_{tt}\delta_{t\tau})$ is a diagonal matrix, then (63.1)-(63.3) is reduced to the conditions

$$B\,1_r = 1_r, B\ge 0 \tag{64.1}$$

$$B\,\Xi_t' = h_{tt}\Xi_t', t=1,\dots,L, \tag{64.2}$$

where 1_r, $\Xi_1',\ldots,\Xi_L'$ are mutually orthogonal r-vectors, see (60.1a), hence $1+L\leq r$. If $u_{L+2},\ldots,u_r$ are additional row vectors such that $\{1_r', \Xi_1,\ldots,\Xi_L, u_{L+2},\ldots,u_r\}$ is an orthogonal basis of $\mathbb{R}^r$, then the symmetric solutions B of (64) have the form

$$B = \frac{1}{(\sqrt{r})^2} 1_r 1_r' + \sum_{t=1}^{L} \frac{h_{tt}}{||\Xi_t||^2} \Xi_t'\Xi_t + \sum_{t=L+2}^{r} \frac{\lambda_t}{||u_t||^2} u_t'u_t, \tag{65}$$

where the eigenvalues $h_{11},\ldots,h_{LL},\lambda_{L+2},\ldots,\lambda_r$ of B still must be selected such that $B\geq 0$, which is obviously possible in every case.

It is easy to see that a similar representation of B holds if $\alpha_j \neq \frac{1}{r}$ for all $j\in R$ and/or $\Xi_1,\ldots,\Xi_L$ are not orthogonal, see §9.

If $h_{tt}=0, t=1,\ldots,L$ and $\lambda_t=0, t=L+2,\ldots,r$, then $B = \frac{1}{r}11'$ and (63.5) has the form

$A^{(t)}y=b^{(t)}$, $t=1,\ldots,L$,

see (63.5.1) and Example 1.

Example 3. If, for a certain subset K of $\{1,2,\ldots,n\}$, the columns $a_k, k\in K$, of $A(\omega)$ are deterministic, see (36.1), then $a_k=a_k^{(0)}$ and $a_k^{(t)}=0$, $t=1,\ldots,L$, for all $k\in K$, where $a_k^{(t)}$ is the kth column of $A^{(t)}$. Hence, in this case, the subvectors $(x_k)_{k\in K}$ and $(y_k)_{k\in K}$ are not involved in (61.2a),(63.5), respectively.

Example 4. If $(A(\omega),b(\omega))$ has deterministic rows (A_i,b_i) for $i\in J$, see (36.2), then $(A_i,b_i) = (A_i^{(0)},b_i^{(0)})$ and $(A_i^{(t)},b_i^{(t)}) = 0$, $t=1,\ldots,L$, for all $i\in J$, where $(A_i^{(t)},b_i^{(t)})$ is the ith row of $(A^{(t)},b^{(t)})$. Hence, (61.2a),(63.5), respectively, are reduced to the $(m-|J|)\cdot L$ linear equations for (y,B)

$$(A_i^{(1)}y-b_i^{(1)},\ldots,A_i^{(L)}y-b_i^{(L)}) = (A_i^{(1)}x-b_i^{(1)},\ldots,A_i^{(L)}x-b_i^{(L)})\Xi B, i\notin J,$$

$$(A_i^{(1)}y-b_i^{(1)},\ldots,A_i^{(L)}y-b_i^{(L)}) = (A_i^{(1)}x-b_i^{(1)},\ldots,A_i^{(L)}x-b_i^{(L)})H, i\notin J,$$

respectively.

Example 5. Suppose that r is even, and that the elements ξ_t^i of Ξ satisfy the relations

$$\xi_t^j = -\xi_t^{j-\frac{r}{2}} \text{ for all } j = \tfrac{r}{2}+1,\dots,r \text{ and } t=1,\dots,L.$$

Hence, Ξ has the form $\Xi = (\Xi_I, -\Xi_I)$, where Ξ_I is an $L \times \frac{r}{2}$ submatrix of Ξ. Then $\Xi B = \Xi_I(B_I - B_{II})$, where $B = \binom{B_I}{B_{II}}$, and (61.2a) has the form

$$(A^{(1)}y-b^{(1)},\dots,A^{(L)}y-b^{(L)})(\Xi_I,-\Xi_I) =$$

$$(A^{(1)}x-b^{(1)},\dots,A^{(L)}x-b^{(L)})\Xi_I(B_I-B_{II}),$$

while (63.3) has the form

$$\Xi_I(B_I-B_{II}) = H(\Xi_I,-\Xi_I).$$

7.2. Stationary points. We recall that according to Definition 4.1 a point $x \in D$ is called stationary relative to D if and only if the relations (3.1)-(3.3), (46), respectively, only have solutions (y,Π), (y,B), respectively, with $y \in D$, of the type $A^j y = A^j x$ for all $j \in R$.

Using $Z_x = (A^1x-b^1,\dots,A^rx-b^r)$, the D-stationarity can also be formulated this way.

Theorem 7.4. A point $x \in D$ is stationary relative to D if and only if for every solution (y,Π), (y,B), respectively, of (3.1)-(3.3), (46), respectively, it is $Z_y=Z_x$, or $h=y-x$ is not a feasible direction for D at x.

Proof. a) Suppose that for every solution (y,B) of (46) it is $Z_y=Z_x$ or $h=y-x$ is not a feasible direction for D at x. If x is not stationary, then (46) has a solution (y,B) with $y \in D$ and $A^jy \neq A^jx$ for at least one $j \in R$. Hence, $Z_y \neq Z_x$ and $h=y-x$ is a feasible direction for D at x, which is a contradiction to the assumptions.

Consequently, x must be stationary in this case.

b) Conversely, suppose now that x is D-stationary and assume that (46) has a solution (y_o,B_o), $y_o \in \mathbb{R}^n$, such that $Z_{y_o} \neq Z_x$ and $h=y_o-x$ is a feasible direction for D at x. Since the tuple (x,I) always satisfies (46) and (46) contains only linear relations with respect to (y,B), we find that $(y,B) = \lambda(y_o,B_o)+(1-\lambda)(x,I)$ also satisfies (46) for every $0<\lambda<1$. Furthermore, since by assumption $Z_{y_o} \neq Z_x$, and $h=y_o-x$ is a feasible direction for D at x, there exists a number $0<\lambda_o<1$ such that $\eta=x+\lambda_o(y_o-x) \in D$ and $Z_\eta=\lambda_o Z_{y_o}+(1-\lambda_o)Z_x \neq Z_x$. Hence, $(\eta,\lambda_o B_o+(1-\lambda_o)I)$ satisfies (46), where $\eta \in D$ and $A^j x \neq A^j x$ for at least one $j \in R$ which contradicts the stationarity of x. This completes the proof of our theorem.

Note. In the above characterization of a D-stationary point we don't have explicitly the constraint $y \in D$ as in the basic Definition 4.1.

By means of the convex polytope $P_x=\text{conv}\{Z_x,Z_xB^{(2)},\ldots,Z_xB^{(N)}\}$ and the embedding $V_D = \{Z_y : y \in D\}$ of D into the affine subspace $V = \{Z_x : x \in \mathbb{R}^n\}$ of $\mathbb{R}^{m\cdot r}$, see § 7.1.4, the D-stationarity can also be described the following way:

<u>Lemma 7.5.</u> A point $x \in D$ is D-stationary if and only if $V_D \cap P_x=\{Z_x\}$.

Proof. For $x \in D$ the intersection $V_D \cap P_x$ always contains Z_x. On the other hand, see (58), we have that

$$V_D \cap P_x = \{Z_y : \text{There is } y \in \mathbb{R}^n \text{ with } Z_y \in V_D \cap P_x\}$$

$$= \{Z_y : \text{There is a tuple } (y,B) \text{ satisfying (46) with } y \in D\}.$$

Since $x \in D$ is D-stationary if and only if for every solution (y,B) of (46) with $y \in D$ it is $A^j y=A^j x$ for all $j \in R$, we now find that the D-stationarity of x can also be characterized by $V_D \cap P_x = \{Z_x\}$.

An immediate consequence of this lemma is that a D_1-stationary point $x \in D_1$ is also D_2-stationary for every $D_2 \subset \mathbb{R}^n$ such that $x \in D_2 \subset D_1$. Especially, every $\mathbb{R}^n$-stationary point x, contained in a set D, is also D-stationary, hence $S_D \supset D \cap S_{\mathbb{R}^n}$.

7.2.1. Examples.

Example 1. Consider a point $x \in D$ such that $A^i x - b^i = z_0$ for all $i \in R$, where z_0 is an arbitrary, but a fixed m-vector. Then $Z_x B = (z_0, z_0, \ldots, z_0) = Z_x$ for every B satisfying (46.1). Hence, (46.3) yields $(A^1 y - b^1, \ldots, A^r y - b^r) = Z_x B = Z_x$ and therefore $A^j y = A^j x$ for every $j \in R$.

This proves the following lemma:

Lemma 7.6. If there is for $x \in D$ a fixed m-vector z_0 such that $A^i x - b^i = z_0$ for every $i \in R$, then x is D-stationary.

Thus, every $x \in D$ such that $A^i x = b^i, i \in R$, is a stationary point relative to D. Especially, if $b(\omega)$ is constant a.s., then the origin 0 is D-stationary, provided that $0 \in D$.

Example 2. Consider a point $x \in D$ satisfying $Z_x B^{(\nu)} = Z_x$ for all $\nu = 1,2,\ldots,N$, where $B^{(1)} = I$, $B^{(2)}, \ldots, B^{(N)}$ are the extreme points of $\mathcal{B}_\alpha$. Because of (57) we then see that

$$Z_y = \sum_{\nu=1}^{N} \sigma_\nu Z_x B^{(\nu)} = Z_x$$

for all $\sigma_1 \geq 0, \ldots, \sigma_N \geq 0$ with $\sum_{\nu=1}^{N} \sigma_\nu = 1$ and therefore $A^j y = A^j x$, $j \in R$, for every solution (y,B) of (46). Hence, x is D-stationary. Thus, we are interested in the points $x \in \mathbb{R}^n$ having $Z_x B^{(\nu)} = Z_x$ for all $\nu = 1,2,\ldots,N$. Considering especially the extreme points $B^{(\nu)}$ of $\mathcal{B}_\alpha$ given in § 7.1.2 by (49), we see that a point x has the above property $Z_x B^{(\nu)} = Z_x, 1 \leq \nu \leq N$, if and only if $z^i = z^j$ for all $1 \leq i < j \leq r$, hence $z^i = z_0$ for all $i \in R$, see Example 1.

7.2.2. Representation (60) of $(A(\omega),b(\omega))$. From Lemma 7.6 follows that every point $x\in D$ with $A^{(t)}x=b^{(t)}$ for all $t=1,\dots,L$ is stationary relative to D.

If rank $\Xi=L$, then, see § 7.1.6.1, $x\in D$ is stationary relative to D if and only if for every solution (y,B) of (46) with $y\in D$ we have that $A^{(t)}y=A^{(t)}x$, $t=0,1,..,L$. Consequently, if rank $\Xi=L$ and x is D-stationary, then for every solution (y,B,H) with $y\in D$ of (63.1)-(63.5) holds $A^{(t)}y = A^{(t)}x, t=0,1,\dots,L$. For L=1 this is also a sufficient condition.

A numerical example. Let L=1, r=2, $\alpha_1=\alpha_2=\frac{1}{2}$ and $\Xi=(1,-1)$.

We now find $B=\begin{pmatrix}\frac{1+\lambda}{2} & \frac{1-\lambda}{2}\\ \frac{1-\lambda}{2} & \frac{1+\lambda}{2}\end{pmatrix}$, $-1\le\lambda\le+1$, see also Example 2 in § 7.1.6.1. Thus in the present case $x\in D$ is D-stationary if and only if the relations

$$A^{(1)}y = A^{(0)}x$$

$$A^{(1)}y-b^{(1)}=\lambda(A^{(1)}x-b^{(1)}) \text{ for any } -1\le\lambda\le+1$$

imply that $A^{(1)}y=A^{(1)}x$, or $h=y-x$ is not a feasible direction for D at x.

Under certain additional assumptions the above Lemma 7.6 can also be reversed.

Lemma 7.7. Let rank $\Xi=L$, and suppose that for every $x\in D$ there exists a solution $y\in D$ of the linear equations

$$A^{(0)}y=A^{(0)}x \text{ and } A^{(t)}y=b^{(t)},\ t=1,2,\dots,L;$$

then x is D-stationary if and only if $A^{(t)}x=b^{(t)}$, $t=1,\dots,L$.

Proof. As mentioned already, if $A^{(t)}x=b^{(t)}$, $t=1,\dots,L$, for a point $x\in D$, then x is D-stationary. Conversely, let $x\in D$ be D-stationary, and assume that $A^{(t)}x\neq b^{(t)}$ for at least one $1\le t\le L$. Consider then the tuple $(y,B,H)=(y,\alpha 1_r',0)$, where $y\in D$ is a vector such that $A^{(0)}y=A^{(0)}x$ and $A^{(t)}y=b^{(t)}, t=1,\dots,L$. Obviously, $B=\alpha 1_r'$

satisfies (46.1),(46.2) and it is $\Xi B = \Xi\alpha 1_r' = 0 = H\Xi$, see (60.1a); furthermore,

$$(A^{(1)}y-b^{(1)},\dots,A^{(L)}y-b^{(L)}) = (A^{(1)}x-b^{(1)},\dots,A^{(L)}x-b^{(L)})H.$$

Hence, (y,B,H) satisfies (63), and therefore (y,B) satisfies (46), where $y \in D$. Since $A^{(t)}y=b^{(t)}$, $t=1,\dots,L$, and $A^{(t)}x \neq b^{(t)}$ for at least one $1 \le t \le L$, we find $A^{(t)}y \neq A^{(t)}x$ for at least one $1 \le t \le L$. Because of rank $\Xi=L$, we now get $A^j y \neq A^j x$ for at least one $j \in R$, which is a contradiction to the stationarity of D. Hence, D-stationarity of x implies under the above assumptions that $A^{(t)}x=b^{(t)}$, $t=1,\dots,L$.

Note. a) Compare Lemma 7.7 with Example 1 in § 7.1.6.1.
b) An interesting situation arises if $\bar{A}=A^{(0)}=0$.

7.2.3. A sufficient condition for stationary points based on mean and covariance of $A(\omega)x-b(\omega)$. Using Theorem 7.3, we can find stationary points in the following way:

Lemma 7.8. If for a given $x \in D$ the relations (59), i.e.

(59.1) $\bar{A}x = \bar{A}y$

(59.2) $\mathrm{var}(A(\cdot)x-b(\cdot)) - \mathrm{var}(A(\cdot)y-b(\cdot))$ is positive semidefinite,

only have solutions $y \in D$ such that $A^j y = A^j x$ for every $j \in R$, then x is D-stationary.

Proof. Assume that x is not stationary relative to D, then there is a solution (y,B) of (46) with $y \in D$ and $A^j y \neq A^j x$ for at least one $j \in R$. Applying Theorem 7.3a, we find that y satisfies also (59) which is now a contradiction to our assumptions. Consequently, x must be stationary.

7.2.4. Stationary points in case of invariant distributions. Hence, suppose that $U_B\hat{\mathbf{A}} = \hat{\mathbf{A}}C$ and $U_B\hat{\mathbf{b}} = \hat{\mathbf{b}}$ for some matrix $B \in \mathcal{B}_\alpha$ and an nxn matrix C. Then we have this lemma:

Lemma 7.9. Assume that (51) holds and let rank $\mathbf{A}=n$. If x is D-stationary, then $Cx=x$ or $h=Cx-x$ is not a feasible direction for D at x.

Proof. According to Theorem 7.2.2, the tuple (y,B) with $y=Cx$ is a solution of (46) for every $x \in \mathbb{R}^n$. If x is D-stationary, by Theorem 7.4 we must have $Cx=y=x$, since rank $\mathbb{A} = n$, or $h=y-x = Cx-x$ is not a feasible direction for D at x.

7.2.5. Further parametric representations of stationary points. Following, we suppose that D is a polyhedral convex subset of $\mathbb{R}^n$. Then $V_D = \{Z_x: x \in D\}$ is a polyhedral convex subset of the affine linear subspace $V=\{Z_x: x \in \mathbb{R}^n\}$ of $\mathbb{R}^{m \cdot r}$. Hence, V_D can always be represented by linear constraints, i.e.

$$V_D = \{W \in \mathbb{R}^{m \cdot r}: \psi_k(W) \leq (=) d_k, k=1,2,\dots,K\}, \tag{66}$$

where $\psi_1, \psi_2, \dots, \psi_K$ are certain linear forms on $\mathbb{R}^{m \cdot r}$ and $d_1, d_2, \dots, d_K$ are given real numbers.

Therefore, for given $x \in D$, W is an element of $V_D \cap P_x$ if and only if (see § 7.1.4)

i) $W = \sum_{\nu=1}^{N} \sigma_\nu Z_x B^{(\nu)}$, $\sum_{\nu=1}^{N} \sigma_\nu = 1$, $\sigma_\nu \geq 0$, $1 \leq \nu \leq N$

ii) $\sum_{\nu=1}^{N} \sigma_\nu \psi_k(Z_x B^{(\nu)}) \leq (=) d_k$, $k=1,2,\dots,K$.

Since $Z_x \in V_D \cap P_x$, it is $\psi_k(Z_x) \leq (=) d_k$, $k=1,2,\dots,K$, for every $x \in D$. Because of Lemma 7.4 and Lemma 7.5, D-stationary points can now be represented as follows.

Theorem 7.5. A point $x \in D$ is stationary relative to D if and only if the system of linear relations

$$\sum_{\nu=1}^{N} \sigma_\nu \psi_k(Z_x B^{(\nu)}) \leq (=) d_k, \quad k=1,2,\dots,K \tag{67.1}$$

$$\sum_{\nu=1}^{N} \sigma_\nu = 1, \quad \sigma_\nu \geq 0, \quad \nu=1,2,\dots,N \tag{67.2}$$

only has solutions $\sigma=(\sigma_1,\dots,\sigma_N)'$ of the type $\sum_{Z_x B^{(\nu)}=Z_x} \sigma_\nu = 1$.

Proof. According to Lemma 7.5 a D-stationary point x is characterized by

$$\{Z_x\} = V_D \cap P_x = \{ \sum_{\nu=1}^{N} \sigma_\nu Z_x B^{(\nu)} : \sum_{\nu=1}^{N} \sigma_\nu \psi_k(Z_x B^{(\nu)}) \leq (=) d_k,$$

$$1 \leq k \leq K, \sum_{\nu=1}^{N} \sigma_\nu = 1, \sigma_\nu \geq 0, 1 \leq \nu \leq N\}.$$

Hence, x is D-stationary if and only if

$$Z_x = \sum_{\nu=1}^{N} \sigma_\nu Z_x B^{(\nu)} = \Big(\sum_{Z_x B^{(\nu)} = Z_x} \sigma_\nu \Big) Z_x + \sum_{Z_x B^{(\nu)} \neq Z_x} \sigma_\nu Z_x B^{(\nu)}$$

for every solution σ of (67). Since Z_x is an extreme point of P_x, see Lemma 7.4, and $\sum_{\nu=1}^{N} \sigma_\nu Z_x B^{(\nu)} \in P_x$ for every σ satisfying (67.2), the above equation holds if and only if $\sum_{Z_x B^{(\nu)} = Z_x} \sigma_\nu = 1$.

Note. The equations $Z_x B^{(\nu)} = Z_x$ for some indices ν, $1 \leq \nu \leq N$, mean that some of the columns z^i, $1 \leq i \leq r$, of Z_x are equal (cf. Example 1a) which corresponds to the set S defined in § 2.7.

The above Theorem 7.5 can also be formulated in the following way.

<u>Corollary 7.2.</u> Let $U \subset \{1,2,\ldots,N\}$ with $1 \in U$ and consider a point $x \in D$ such that $Z_x B^{(\nu)} = Z_x$ for all $\nu \in U$ and $Z_x B^{(\nu)} \neq Z_x$ for $\nu \notin U$, then x is D-stationary if and only if the linear program

$$\text{minimize } g(\sigma) \quad \text{s.t.} \quad (67), \tag{68}$$

with the objective function $g(\sigma) = \sum_{\nu \in U} \sigma_\nu$, has the optimal value $g^* = 1$.

Proof. From our assumptions we obtain $\sum_{Z_x B^{(\nu)} = Z_x} \sigma_\nu = \sum_{\nu \in U} \sigma_\nu = g(\sigma)$.

Thus, according to Theorem 7.5, $x \in D$ is D-stationary if and only if the linear program (68) has only feasible solutions σ of the type $g(\sigma) = 1$ which yields the assertion.

If V_D (see (66)) is an affine subspace of $\mathbb{R}^{m \cdot r}$, we then have this next corollary:

Corollary 7.3. Let V_D be defined by $V_D=\{W\in\mathbb{R}^{m\cdot r}$: $\psi_k(W)=d_k,\ k=1,2,\ldots,K\}$. If, for an $x\in D$, there exists an integer $1\le k\le K$ and a subset $U\subset\{1,2,\ldots,N\}$ with $1\in U$ such that $Z_xB^{(\nu)}=Z_x$ for all $\nu\in U$ and

$$\psi_k(Z_x B^{(\nu)})>d_k\quad(\psi_k(Z_xB^{(\nu)})<d_k,\ \text{resp.})\quad\text{for all } \nu\notin U, \tag{69}$$

then x is D-stationary.

Proof. First, we note that $\psi_k(Z_x)=d_k$, $k=1,\ldots,K$, and $U=\{\nu:\ Z_x B^{(\nu)}=Z_x\}$. For every element $W=\sum_{\nu=1}^{N}\sigma_\nu Z_xB^{(\nu)}$ of P_x it is $\psi_k(W)=\sum_{\nu=1}^{N}\sigma_\nu\ \psi_k(Z_xB^{(\nu)})=\sum_{\nu\notin U}\sigma_\nu\ \psi_k(Z_xB^{(\nu)})+(\sum_{\nu\in U}\sigma_\nu)d_k$.

Hence, $\psi_k(W)>d_k(\psi_k(W)<d_k$, resp.) if $\sigma_\nu>0$ for some $\nu\notin U$.

Consequently, the relations (67) can only have solutions σ of the type $\sum_{Z_xB^{(\nu)}=Z_x}\sigma_\nu=\sum_{\nu\in U}\sigma_\nu=1$. The assertion follows now from Theorem 7.5.

Note. Condition (69) means that all matrices $Z_xB^{(\nu)}$, different from Z_x, lie above (below resp.) the hyperplane $\psi_k(W)=d_k$ in $\mathbb{R}^{m\cdot r}$. It is easy to see that for K=1 the above criterion is also necessary for the stationarity of x.

If V_D is an affine linear space, then the linear program (68) can be modified as follows.

Corollary 7.4. Let V_D be an affine linear space. A vector $x\in D$ such that $Z_xB^{(\nu)}=Z_x$ for $\nu\in U$ and $Z_xB^{(\nu)}\neq Z_x$ for $\nu\notin U$, where U is a subset of $\{1,2,\ldots,N\}$ with $1\in U$, is stationary if and only if the linear program

$$\text{maximize } h(\sigma) \tag{70}$$

$$\text{s.t.}\ \sum_{\nu\notin U}\sigma_\nu(\psi_k(Z_xB^{(\nu)})-d_k)=0,\ k=1,2,\ldots,K \tag{70.1}$$

$$\sigma_\nu\ge 0 \text{ for } \nu\notin U, \tag{70.2}$$

having the objective function $h(\sigma) = \sum_{\nu \notin U} \sigma_\nu$ has the optimal value $h^*=0$.

Proof. Consider a point $x \in D$ such that $Z_x B^{(\nu)} = Z_x$ for $\nu \in U$ and $Z_x B^{(\nu)} \neq Z_x$ for $\nu \notin U$. If V_D is affine, then relation (67) has the form

$$\sum_{\nu \notin U} \sigma_\nu \psi_k(Z_x B^{(\nu)}) = (\sum_{\nu \notin U} \sigma_\nu) d_k, \quad k=1,\ldots,K \tag{71.1}$$

$$\sum_{\nu \notin U} \sigma_\nu + \sum_{\nu \in U} \sigma_\nu = 1, \quad \sigma_\nu \geq 0, \quad \nu=1,\ldots,N. \tag{71.2}$$

According to Theorem 7.5 a point $x \in D$ is stationary relative to D if and only if (71) only has solutions σ of the type $\sigma_\nu=0$ for all $\nu \notin U$. Hence, x is D-stationary if and only if the linear program (70) has the unique optimal solution $\sigma_\nu^*=0$ for all $\nu \notin U$. This proves our corollary.

Note. a) If $U = \{1,2,\ldots,N\}$, then we set $\sum_{\nu \notin U} \sigma_\nu = 0$.

b) According to Gordan's theorem [54], the linear program (70) has a nonzero feasible solution $(\sigma_\nu)_{\nu \notin U}$ if and only if the system of linear inequalities

$$\sum_{k=1}^{K} w_k(\psi_k(Z_x B^{(\nu)}) - d_k) < 0 \text{ for all } \nu \notin U$$

has no solution $w = (w_1,\ldots,w_K)' \in \mathbb{R}^K$. This means for K=1 that there exist integers $1 \leq \nu_1, \nu_2 \leq N$ such that the corresponding numbers $\psi_1(Z_x B^{(\nu_1)})$, $\psi_1(Z_x B^{(\nu_2)})$ have opposite signs, cf. Corollary 7.3.

c) For a set $U \subset \{1,2,\ldots,N\}$ with $1 \in U$ let M_U be defined by

$$M_U = \{x \in D: Z_x B^{(\nu)} = Z_x \text{ for all } \nu \in U \text{ and (67) has only solutions } \sigma=(\sigma_1,\sigma_2,\ldots,\sigma_N)' \text{ such that } \sum_{\nu \in U} \sigma_\nu = 1\}.$$

Concerning the set S_D of all D-stationary points we then have also this result:

Corollary 7.5. $S_D = \{x: x \in M_U$ for some $U \subset \{1,2,...,N\}$ with $1 \in U\}$.

Proof. Let $x \in M_U$, where $1 \in U \subset \{1,2,...,N\}$. According to the definition of M_U, for every solution σ of (67) we find

$1 \geq \sum_{Z_x B^{(\nu)} = Z_x} \sigma_\nu \geq \sum_{\nu \in U} \sigma_\nu = 1$, hence x is D-stationary by Theorem 7.5. On the other hand, if x is D-stationary, then $x \in M_{U_x}$, where U_x is defined by

$$U_x = \{\nu: 1 \leq \nu \leq N,\ Z_x B^{(\nu)} = Z_x\}.$$

Note. Because of Corollary 7.5 for S_D we also have the representation

$$S_D = \{x \in D: x \in M_{U_x}\}.$$

Obviously, we may arrange the constraints in (66) such that the linear program (68) can be respresented by

$$\text{minimize } g_U'\sigma \tag{72}$$

$$\text{s.t. } \begin{pmatrix} p_1^1(x)-d_1 & p_1^2(x)-d_1 & \cdots & p_1^N(x)-d_1 \\ p_2^1(x)-d_2 & p_2^2(x)-d_2 & \cdots & p_2^N(x)-d_2 \\ \vdots & \vdots & & \vdots \\ p_L^1(x)-d_L & p_L^2(x)-d_L & \cdots & p_L^N(x)-d_L \\ 1 & 1 & \cdots & 1 \end{pmatrix} \sigma = \begin{pmatrix} 0 \\ 0 \\ \vdots \\ 0 \\ 1 \end{pmatrix}$$

$$\begin{pmatrix} p_{L+1}^1(x)-d_{L+1} & p_{L+1}^2(x)-d_{L+1} & \cdots & p_{L+1}^N(x)-d_{L+1} \\ \vdots & \vdots & & \vdots \\ p_K^1(x)-d_K & p_K^2(x)-d_K & \cdots & p_K^N(x)-d_K \end{pmatrix} \sigma \leq \begin{pmatrix} 0 \\ \vdots \\ 0 \end{pmatrix}$$

$$\sigma \geq 0,$$

where L, $0 \leq L \leq K$ is the number of equality constraints in (66), the N-vector g_U is defined by

$$g_{U\nu} = 1 \text{ for } \nu \in U \text{ and } g_{U\nu} = 0 \text{ for } \nu \notin U,$$

furthermore, $p_k^\nu(x)$ is given by

$$p_k^\nu(x) = \psi_k(Z_x B^{(\nu)}),\ \nu=1,2,\dots,N,\ k=1,2,\dots,K.$$

If $x \in D$, then the linear program (68) has always the feasible solution $\sigma=(1,0,\dots,0)'$, and the objective function of (68) is bounded from below on the feasible domain of (68). Hence, (68) has an optimal solution $\sigma^*=\sigma^*(x)$ for each $x \in D$. Since (72) is only another representation of the LP (68), for every $x \in D$ we find that the optimal value $g_U^*(x)$ (≤ 1) of (68) is equal to the optimal value of the dual linear program of (72)

$$\begin{aligned} &\text{maximize } w \\ \text{s.t. } &(p^\nu(x)-d)'\binom{u}{-v} + w \leq g_{U\nu},\ \nu=1,2,\dots,N \\ &u \in \mathbb{R}^L,\ v \in \mathbb{R}^{K-L},\ v \geq 0, \end{aligned} \tag{73}$$

where

$$p^\nu(x) = (\psi_1(Z_x B^{(\nu)}),\ \psi_2(Z_x B^{(\nu)}),\dots,\psi_K(Z_x B^{(\nu)}))',$$

$$d = (d_1,d_2,\dots,d_K)'.$$

Because of the duality between (68) and (73), for each $x \in D$ we find that

$$g_U^*(x) = \max\{\phi_U(x;u,v):\ u \in \mathbb{R}^L,\ v \in \mathbb{R}^{K-L},\ v \geq 0\}, \tag{74.1}$$

where $\phi_U(x;u,V)$ is defined by

$$\phi_U(x;u,v) = \min_{1 \leq \nu \leq N} g_{U\nu} - (p^\nu(x)-d)'\binom{u}{-v}. \tag{74.2}$$

For $x \in \mathbb{R}^n$ let

$$U_x = \{\nu:\ 1 \leq \nu \leq N,\ Z_x B^{(\nu)} = Z_x\};$$

moreover, define

$p_I^\nu(x) = (\psi_1(Z_x B^{(\nu)}),\ldots,\psi_L(Z_x B^{(\nu)}))'$, $d_I=(d_1,\ldots,d_L)'$,

$p_{II}^\nu(x) = (\psi_{L+1}(Z_x B^{(\nu)}),\ldots,\psi_K(Z_x B^{(\nu)}))'$, $d_{II}=(d_{L+1},\ldots,d_K)'$,

If $x \in D$ and $\nu \in U_x$, then

$$p_k^\nu(x) = \psi_k(Z_x B^{(\nu)}) = \psi_k(Z_x) = d_k,\ k=1,\ldots,L,$$

$$p_k^\nu(x) = \psi_k(Z_x B^{(\nu)}) = \psi_k(Z_x) \le d_k,\ k=L+1,\ldots,K.$$

Hence, if $x \in D$, then

$$\phi_{U_x}(x;u,v) = \min\{1+(p_{II}^1(x)-d_{II})'v,\ -(p_I^\nu(x)-d_I)'u+(p_{II}^\nu(x)-d_{II})'v, \nu \notin U_x\}. \tag{75}$$

Note that $p_{II}^1(x)-d_{II} \le 0$ for all $x \in D$.

This representation (75) of $\phi_{U_x}(x;u,v)$ yields the next lemma.

Lemma 7.10. a) If, for a $x \in D$, there is an integer ν, $1<\nu\le N$, such that $Z_x B^{(\nu)} \neq Z_x$ and

$$p_I^\nu(x) = d_I \text{ and } p_{II}^\nu(x) \le d_{II}, \tag{76}$$

then x is not D-stationary. b) If $x \in D$ is D-stationary and (76) holds for some ν, $1<\nu\le N$, then $Z_x B^{(\nu)}=Z_x$.

Proof. a) Let $x \in D$ and $1<\nu\le N$ be an integer such that $Z_x B^{(\nu)} \neq Z_x$ and (76) holds. By (75) we find

$$\phi_{U_x}(x;u,v) \le -(p_I^\nu(x)-d_I)'u+(p_{II}^\nu(x)-d_{II})'v = (p_{II}^\nu(x)-d_{II})'v \le 0$$

for every $u \in \mathbb{R}^L$ and $v \in \mathbb{R}^{K-L}$, $v\ge 0$. Consequently, (74) yields $g^*_{U_x}(x) \le 0$, and the assertion follows now from Corollary 7.2.
b) If x is D-stationary, and (76) holds for some $1<\nu\le N$, then $Z_x B^{(\nu)}=Z_x$, since otherwise we get a contradiction by means of the first part of this lemma.

Note. Condition (76) means that $Z_x B^{(\nu)} \in V_D$, i.e. there is an element $\tilde{x} \in D$ such that

$$Z_x B^{(\nu)} = Z_{\tilde{x}}.$$

An important consequence of Lemma 7.10b is contained in

Corollary 7.6. Suppose that there is a fixed integer ν, $1<\nu\leq N$, such that (76) holds for every $x\in D$, i.e. $Z_x B^{(\nu)}\in V_D$ for each $x\in D$, then it is $Z_x B^{(\nu)} = Z_x$ for every D-stationary point $x\in D$.

According to § 7.2.1, Example 2, we know that a point $x\in D$, fulfilling

$$U_x = \{\nu: Z_x B^{(\nu)} = Z_x\} = \{1,2,\dots,N\},$$

is D-stationary. Because of Corollary 7.2 and (74), (75) the D-stationary points x can now be characterized as follows.

Theorem 7.6. A point $x\in D$ is D-stationary if and only if there exist vectors $u\in\mathbb{R}^L$ and $v\in\mathbb{R}^{K-L}$, $v\geq 0$ such that

$$(p^1_{II}(x) - d_{II})'v = 0, \tag{77.1}$$

$$-(p^\nu_I(x)-d_I)'u + (p^\nu_{II}(x)-d_{II})'v\geq 1 \text{ for all } \nu\notin U_x. \tag{77.2}$$

Proof. Let $x\in D$. If $U_x = \{1,2,\dots,N\}$, then, by the above remark, we know that x is D-stationary; furthermore, (77.2) is then cancelled and (77.1) holds with v=0. Hence, consider now the case $U_x \neq \{1,2,\dots,N\}$. By (74) we know that $g^*_{U_x}(x) = \phi_{U_x}(x;u^*,v^*)$ for some vectors $u^*\in\mathbb{R}^L$, $v^*\in\mathbb{R}^{K-L}$, $v^*\geq 0$, where $\phi_{U_x}(x;u^*,v^*)$ is given by (75). Hence, $x\in D$ satisfies $g^*_{U_x}(x) = 1$ if and only if there are vectors $u\in\mathbb{R}^L$, $v\in\mathbb{R}^{K-L}$, $v\geq 0$, such that

$$1 + (p^1_{II}(x) - d_{II})'v\geq 1, \tag{78.1}$$

$$-(p^\nu_I(x)-d_I)'u+(p^\nu_{II}(x)-d_{II})'v\geq 1 \text{ for all } \nu\notin U, \tag{78.2}$$

where the equality sign holds at least once in (78). If $0\leq L<K$, then $p^1_{II}(x)-d_{II}\leq 0$. Hence, because of $v\geq 0$, (78.1) holds if and only if $(p^1_{II}(x)-d_{II})'v = 0$. Since (78.1), in this case, is always an equality constraint, the D-stationary points are characterized here in fact by (77). Now, we consider the case L=K. Then, (78) is

reduced to

$$-(p_I^\nu(x)-d_I)'u \geq 1 \text{ for all } \nu \notin U_x, \tag{79.1}$$

$$-(p_I^\nu(x)-d_I)'u = 1 \text{ for at least one } \nu \notin U_x; \tag{79.2}$$

furthermore, (77) has then the form

$$-(p_I^\nu(x)-d_I)'u \geq 1 \text{ for all } \nu \notin U_x.$$

Obviously, (79) implies (77) in the present case L=K. Conversely, assume that (77) holds for some $u \in \mathbb{R}^K$. If the equality sign holds at least once in (77), then (79) is also fulfilled, and the proof is complete. Otherwise, if $-(p_I^\nu(x)-d_I)'u>1$ for all $\nu \notin U_x$, then we consider $\tilde{u} = \lambda u$ with $\lambda \in \mathbb{R}$. Selecting λ such that $0<\lambda<1$ and

$$\lambda \cdot \min_{\nu \notin U_x} (-(p_I^\nu(x)-d_I)'u = 1,$$

we find $-(p_I^\nu(x)-d_I)'\tilde{u} \geq 1$ for all $\nu \notin U_x$ and $-(p_I^\nu(x)-d_I)'\tilde{u} = 1$ for at least one $\nu \notin U_x$. Hence, (79) holds with $u:=\tilde{u}$ which now concludes the proof of our lemma.

Note. a) Theorem 7.6 yields a parametric representation of the set S_D of D-stationary points.

b) It is easy to see that (77) can be replaced by the equivalent condition

$$(p_{II}^1(x)-d_{II})'v = 0, \tag{80.1}$$

$$-(p_I^\nu(x)-d_I)'u + (p_{II}^\nu(x)-d_{II})'v > 0 \text{ for all } \nu \notin U_x. \tag{80.2}$$

c) It is easy to see that the vector $p^\nu(x) = (\psi_1(Z_x B^{(\nu)}),\ldots, \psi_K(Z_x B^{(\nu)}))'$ has the form

$$p^\nu(x) = P^\nu x - q^\nu; \tag{81}$$

if b_{ij}^ν, $i,j=1,\ldots,r$, denote the elements of $B^{(\nu)}$, and Ψ is the $K \times (r\cdot m)$ matrix representing the linear forms $\psi_1,\ldots,\psi_K$, then the $K \times n$ matrix P^ν and the K-vector q^ν, $\nu=1,2,\ldots,N$, are given by

$$P^{\nu} = \Psi \begin{pmatrix} \sum_{i=1}^{r} b_{i1}^{\nu} A^{i} \\ \vdots \\ \sum_{i=1}^{r} b_{ir}^{\nu} A^{i} \end{pmatrix}, \quad q^{\nu} = \Psi \begin{pmatrix} \sum_{i=1}^{r} b_{i1} b^{i} \\ \vdots \\ \sum_{i=1}^{r} b_{ir}^{\nu} b^{i} \end{pmatrix}. \qquad (81.1)$$

If $\alpha_j = \frac{1}{r}$ for all $j=1,\dots,r$, then $B^{(\nu)}$, $\nu=1,\dots,N=r!$, are the permutation matrices of $\{1,2,\dots,r\}$, and we have

$$P^{\nu} = \Psi \begin{pmatrix} A^{\tilde{\nu}(1)} \\ \vdots \\ A^{\tilde{\nu}(r)} \end{pmatrix}, \qquad q^{\nu} = \Psi \begin{pmatrix} b^{\tilde{\nu}(1)} \\ \vdots \\ b^{\tilde{\nu}(r)} \end{pmatrix},$$

where $\tilde{\nu}$ denotes the ν-th permutation of $\{1,2,\dots,r\}$.

There are two main special situations to be considered now.

A) Let $L=K$, i.e. there are only equality constraints in (66). Thus, $p_I^{\nu}(x)-d_I=p^{\nu}(x)-d=P^{\nu}x-(q^{\nu}+d)$ and condition (80) has the form

$$(p^{\nu}(x)-d)'u<0 \text{ for all } \nu \notin U_x. \qquad (82)$$

According to Theorem 7.6 a point $x \in D$ is D-stationary in the present case if and only if $U_x = \{1,2,\dots,N\}$, or there is a vector $u \in \mathbb{R}^K$ such that (82) holds. Consequently, if $x \in D$ is D-stationary, then

$$p^{\nu}(x) = d \text{ if and only if } \nu \in U_x, \qquad (83)$$

cf. Lemma 7.10. Hence, if $L=K$, then we also have this characterization of D-stationary points.

Lemma 7.11.1. Let $L=K$ and $x \in D$. a) If x is D-stationary, then (83) holds and $0(=p^1(x)-d)$ is an extreme point of $\text{conv}\{p^{\nu}(x)-d: \nu=1,2,\dots,N\}$. b) If x fulfills (83), and 0 is an extreme point of $\text{conv}\{p^{\nu}(x)-d: \nu=1,2,\dots,N\}$, then x is D-stationary.

Proof. It is $p^{\nu}(x)-d=0$ for all $\nu \in U_x$, hence, the assertion obviously holds if $U_x = \{1,2,\dots,N\}$. Let now be $U_x \neq \{1,2,\dots,N\}$.

a) Suppose that x is D-stationary, then (83) holds, and there is a vector $u \in \mathbb{R}^K$ such that (82) holds. If 0 is not an extreme point of $C = \text{conv}\{p^\nu(x)-d: \nu=1,2,\dots,N\}$, then, because of (83), there are numbers $\sigma_\nu \geq 0$, $\nu \notin U_x$, $\sum_{\nu \notin U_x} \sigma_\nu = 1$ such that

$0 = \sum_{\nu \notin U_x} \sigma_\nu (p^\nu(x)-d)$, which is a contradiction to (82). Hence, 0 must be an extreme point of C. b) Conversely, suppose $x \in D$ is a point such that (83) holds, and 0 is an extreme point of C. Assuming that (82) has no solution $u \in \mathbb{R}^K$, using Gordan's theorem [46], we know that there are numbers $\tau_\nu \geq 0$, $\nu \notin U_x$, $\sum_{\nu \notin U_x} \tau_\nu > 0$,

such that $0 = \sum_{\nu \notin U_x} \tau_\nu (p^\nu(x)-d)$. However, this is a contradiction to the supposition that 0 is an extreme point of C. Hence, (82) has a solution $u \in \mathbb{R}^K$ which proves that x is D-stationary.

The above considerations yield the following consequences.

A1) Let L=K and let $x \in D$ be any point with $U_x \neq \{1,2,\dots,N\}$. Given numbers $\sigma_\nu \geq 0$, $\nu \notin U_x$, $\sum_{\nu \notin U_x} \sigma_\nu > 0$, assume that x satisfies the equation

$$\Big(\sum_{\nu \notin U_x} \sigma_\nu P^\nu\Big)x - \sum_{\nu \notin U_x} \sigma_\nu (q^\nu + d) = 0, \tag{84.1}$$

then x is not D-stationary. This is an immediate consequence of Lemma 7.11.1 and (81).

A2) Let L=K and $u \in \mathbb{R}^K$, $u \neq 0$. Every vector $x \in D$ satisfying

$$(p^\nu(x)-d)'u < 0 \text{ for all } \nu=2,3,\dots,N \tag{84.2}$$

is a D-stationary point.

A3) Let L=K and $\bar{\nu}$ an integer with $1 < \bar{\nu} \leq N$. Given positive numbers β_ν, $\nu=2,\dots,N$, where $\beta_{\bar{\nu}}=1$, suppose that $x \in D$ fulfills the relations

$$p^{\bar{\nu}}(x)-d \neq 0,$$
$$p^{\nu}(x)-d = \beta_{\nu}(p^{\bar{\nu}}(x)-d) \text{ for all } \nu=2,\dots,N, \tag{84.3}$$

then x is D-stationary. Indeed, choosing a vector $u \in \mathbb{R}^K$ such that $(p^{\bar{\nu}}(x)-d)'u<0$, because of (84.3), we have $(p^{\nu}(x)-d)'u = \beta_{\nu}(p^{\bar{\nu}}(x)-d)'u<0$ for all $\nu \neq 1$. Hence, x is D-stationary.

A4) Let L=K. Suppose that D is a linear space and $b^i=0$ for all $i \in R$. Then V_D is a linear space, hence $d_k=0$ for all $k=1,2,\dots,K$, and $q^{\nu}=0$, $\nu=1,2,\dots,N$. Consequently, $p^{\nu}(x)-d = P^{\nu}x$, and (82) has the form

$$(P^{\nu\prime}u)'x < 0 \text{ for all } \nu \notin U_x. \tag{84.4}$$

Obviously, 0 is D-stationary. Since $Z_{\lambda x} = \lambda Z_x$, we have $U_{\lambda x} = U_x$ for all $\lambda \neq 0$ and for a D-stationary point $x \in D$, λx is also D-stationary for every $\lambda \in \mathbb{R}$.

A5) Let L=K=1. According to Theorem 7.6. and (82) a point $x \in D$ is D-stationary if and only if either

$$p^{\nu}(x)-d = P^{\nu}x-(q^{\nu}+d)<0 \text{ for all } \nu \notin U_x$$

or (84.5)

$$p^{\nu}(x)-d = P^{\nu}x-(q^{\nu}+d)>0 \text{ for all } \nu \notin U_x.$$

Conversely, if there are for a vector x integers $1<\mu, \nu \leq N$ such that

$$p^{\nu}(x)-d < 0 < p^{\mu}(x)-d,$$

then x is not D-stationary.

Note. Corollary 7.3 follows from Theorem 7.6 by choosing $u = (0,\dots,0,u_k,0,\dots,0)$, $u_k \neq 0$.

B) Let L=0, i.e. there are only inequality constraints in (66). Then $p^{\nu}_{II}(x)-d_{II} = p^{\nu}(x)-d = P^{\nu}x-(q^{\nu}+d)$, and (80) has the form

$(p^1(x)-d)'v = 0,$ (85.1)

$(p^\nu(x)-d)'v > 0$ for all $\nu \notin U_x$. (85.2)

According to Theorem 7.6 a point $x \in D$ is here D-stationary if and only if there is a vector $v \in \mathbb{R}^K$ with $v \geq 0$ such that (85) holds. Hence, if x is D-stationary, then

$p^\nu(x) \leq d$ if and only if $\nu \in U_x$, (86)

cf. (83). Corresponding to Lemma 7.11.1 we find here this lemma:

Lemma 7.11.2. Let L=0. If $x \in D$ is D-stationary, then (86) holds, and $p^1(x)-d$ is an extreme point of $\mathrm{conv}\{p^\nu(x)-d: \nu=1,2,...,N\}$.

Proof. It is $p^\nu(x)-d=p^1(x)-d$ for all $\nu \in U_x$, hence the assertion obviously holds if $U_x = \{1,2,...,N\}$. Take now $U_x \neq \{1,2,...,N\}$. If x is D-stationary, then we know that (86) holds, and that there is a vector $v \geq 0$ such that (85) is fulfilled. Assuming then that $p^1(x)-d$ is not an extreme point of $C = \mathrm{conv}\{p^\nu(x)-d: \nu=1,2,...,N\}$, there are numbers $\sigma_\nu \geq 0$, $\nu \notin U_x$, $\sum_{\nu \notin U_x} \sigma_\nu = 1$ such that $p^1(x)-d = \sum_{\nu \notin U_x} \sigma_\nu (p^\nu(x)-d)$. By means of (85) we then obtain the contradiction $0 = (p^1(x)-d)'v = \sum_{\nu \notin U_x} \sigma_\nu (p^\nu(x)-d)'v > 0$. Hence, $p^1(x)-d$ is an extreme point of C.

An important property of D-stationary points is given in the next lemma:

Lemma 7.12. If L=0 and $x \in D$ is a D-stationary point such that $U_x \neq \{1,2,...,N\}$, then $Z_x = (A^1x-b^1,...,A^rx-b^r)$ lies at least on one supporting hyperplane $\{W: \psi_K(W)=d_k\}$ of V_D.

Proof. Let $x \in D$ be a D-stationary point with $U_x \neq \{1,2,...,N\}$. Hence, there exists a vector $v \geq 0$ satisfying (85.1) and (85.2). Because of $v \geq 0$ and (85.2), there are components v_k, $k=k_1,...,k_\kappa$, of v such that $v_k > 0$, $k=k_1,...,k_\kappa$. By (85.1) this yields

$\psi_k(Z_x)-d_k=p_k^1(x)-d_k=0$ for $k=k_1,\dots,k_\kappa$. Since $V_D=\{W\in\mathbb{R}^{m\cdot r}: \psi_k(W)\le d_k,\ k=1,\dots,K\}$, this means that Z_x lies on the supporting hyperplanes $\{W\in\mathbb{R}^{m\cdot r}: \psi_k(W)=d_k\}$, $k=k_1,\dots,k_\kappa$, of V_D.

We now note some consequences of the above considerations.

B1) Let L=0 and consider any point $x\in D$ with $U_x\neq\{1,2,\dots,N\}$. Given arbitrary numbers $\sigma_\nu\ge 0$, $\nu\notin U_x$, $\sum_{\nu\notin U_x}\sigma_\nu=1$, assume that x satisfies the equation

$$p^1(x)-d = \sum_{\nu\notin U_x}\sigma_\nu(p^\nu(x)-d), \tag{87.1}$$

then x is not D-stationary. This is an immediate consequence of Lemma 7.11.2.

B2) Let L=0 and $v\in\mathbb{R}_+^K$, $v\neq 0$. Every vector $x\in D$ satisfying

$$\begin{aligned}&(p^1(x)-d)'v = 0\\ &(p^\nu(x)-d)'v > 0 \text{ for all } \nu=2,3,\dots,N\end{aligned} \tag{87.2}$$

is D-stationary.

B3) Let L=0 and K=1. According to Theorem 7.6 and (85) a point $x\in D$ having $U_x\neq\{1,2,\dots,N\}$ is D-stationary in this case if and only if

$$\begin{aligned}&p^1(x)-d = P^1x - (q^1+d) = 0\\ &p^\nu(x)-d = P^\nu x - (q^\nu+d) > 0 \text{ for all } \nu\notin U_x.\end{aligned} \tag{87.3}$$

Note that in the present case $p^\nu(x) = \psi_1(Z_xB^{(\nu)})$, $d=d_1$ and $V_D = \{W\in\mathbb{R}^{m\cdot r}: \psi_1(W)\le d_1\}$. Hence, if x is D-stationary and $U_x \neq \{1,2,\dots,N\}$, then Z_x lies on the supporting hyperplane $\{W\in\mathbb{R}^{m\cdot r}: \psi_1(W)=d_1\}$ of the half space V_D, cf. Lemma 7.12.

7.3. Stochastic optimization problems with a non-strictly convex loss function u. If the convex loss function u of our (SOP) is not strictly convex, see e.g. the class of stochastic linear programs with recourse described in § 2.6, then it may happen that the objective function $F(x) = Eu(A(\omega)x-b(\omega))$, $x\in\mathbb{R}^n$, of

(SOP), see (1), is constant on certain line segments $\overline{xy}$, though $A^j y \neq A^j x$ for at least one $j \in R$. Hence, the necessary optimality condition in Lemma 4.1, that an optimal solution x^* of (SOP) is a D-stationary point, can not be applied immediately. Replacing, therefore, the convex loss function u by the strictly convex function

$$u_\rho(z) = u(z) + \rho||z||^2, \; z \in \mathbb{R}^m,$$

where $||\cdot||$ denotes the Euclidean norm and $\rho>0$ is a (small) positive parameter, the objective function F is substituted by the convex function

$$F_\rho(x) = Eu_\rho(A(\omega)x-b(\omega)) = F(x) + \rho E||A(\omega)x-b(\omega)||^2.$$

Suppose that there is a positive number M such that

$$||A^i x-b^i|| \leq M \text{ for all } x \in D \text{ and all } i \in R, \qquad (88)$$

then for every $x \in D$ and $\rho>0$ we obtain

$$0 \leq F_\rho(x)-F(x) \leq \rho E||A(\omega)x-b(\omega)||^2 \leq \rho M^2.$$

Assuming that $F^* = \inf\{F(x): x \in D\}>-\infty$, the above inequalities yield

$$0 \leq F_\rho^* - F^* \leq \rho M^2,$$

where $F_\rho^* = \inf\{F_\rho(x): x \in D\}$. Thus $\lim_{\rho \downarrow 0} F_\rho^* = F^*$. Furthermore, if x_ρ^* is an optimal solution of the approximative minimization problem minimize $F_\rho(x)$ s.t. $x \in D$ and $x^* \in D$ is an accumulation point of the familiy $(x_\rho^*)_{\rho>0}$ as $\rho \downarrow 0$, then x^* is an optimal solution of (SOP). Concerning the approximative solutions x_ρ^*, $\rho>0$, of (SOP) we have this decisive lemma:

<u>Lemma 7.13.</u> If x_ρ^* is an optimal solution of the approximative (SOP) minimize $F_\rho(x)$ s.t. $x \in D$, then x_ρ^* is a D-stationary point according to Definition 4.1.

Proof. Suppose that x_ρ^* is not D-stationary, then there exists a solution (y,B) of (46), where $x = x_\rho^*$, such that $y \in D$ and $A^j y \neq A^j x_\rho^*$ for at least one $j \in R$. This yields $F_\rho(y) \le F_\rho(x^*)$, and because of the strict convexity of u_ρ it is $F_\rho(\lambda y+(1-\lambda)x_\rho^*) < \lambda F_\rho(y) + (1-\lambda)F_\rho(x_\rho^*) \le F_\rho(x_\rho^*)$ for every $0<\lambda<1$. Hence, $h=y-x_\rho^*$ is a feasible descent direction for F_ρ at x_ρ^* which is a contradiction to the optimality of x_ρ^*. Thus, an optimal solution x_ρ^* of $\min F_\rho(x)$ s.t. $x \in D$ is a D-stationary point.

From the above considerations we obtain now this theorem:

Theorem 7.7. Let D be a compact convex subset of $\mathbb{R}^n$. Then there exists at least one optimal solution x^* of (SOP) lying in the closure $\bar{S}_D$ of the set S_D of D-stationary points.

Proof. Since D is compact, (SOP) has an optimal solution, and also each approximative problem $\min_{x\in D} F_\rho(x)$ has an optimal solution x_ρ^* for every $\rho>0$; furthermore, $F^*>-\infty$ and condition (88) is fulfilled. According to Lemma 7.13 we know that $x_\rho^* \in S_D$ for every $\rho>0$. Because of the compactness of D, there exists an accumulation point $x^* \in D$ of $(x_\rho^*)_{\rho>0}$ as $\rho \downarrow 0$. It is $x^* \in \bar{S}_D$, and by the preceding considerations we know that x^* is an optimal solution of (SOP) which now concludes the proof of our theorem.

Note. a) Theorem 7.7 holds also if one simply assumes that D is a closed, convex set, $\min_{x\in D} F_\rho(x)$ has an optimal solution for every $\rho>0$ and $\lim_{\rho\downarrow 0} F_\rho^* = F^*>-\infty$.

b) In many cases, one has $\bar{S}_D = S_D$. The closure $\bar{S}_D$ of S_D is studied later on.

7.3.1. An application: Stochastic linear programs with recourse. According to § 2.6 we know, that in this case, the objective function F is given by

$$F(x) = \bar{c}_0'x + Ep(A_0(\omega)x-b_0(\omega)),$$

where p is a sublinear function. Hence, it may happen that F is constant on certain line segments xy, though $A_o^j y \neq A_o^j x$ for at least one $j \in R$. However, according to Theorem 7.7. we know that we may also work in this situation essentially with the D-stationary concept given by Definition 4.1.

Since stochastic linear programs with recourse have a partly monotone loss function u, the relationship between $P_{A(\cdot)x-b(\cdot)}$ and $P_{A(\cdot)y-b(\cdot)}$ can be described more adequately by the system of relations (10). Consequently, in the present case Definition 4.1 may be replaced by this sharper definition:

Definition 7.1. In a stochastic linear program with recourse a point $x \in D$ is called (SLP)-D-stationary if the system of relations

$$
\begin{aligned}
&1_r{}'B = 1_r{}', \ B \geq 0, \\
&B\alpha = \alpha \\
&\bar{c}_o{}'y \leq \bar{c}_o{}'x \\
&(A_o^1 y - b_o^1, \ldots, A_o^r y - b_o^r) = (A_o^1 x - b_o^1, \ldots, A_o^r x - b_o^r)B \\
&y \in D
\end{aligned}
\qquad (89)
$$

only has solutions (y,B) such that $\bar{c}_o{}'y = \bar{c}_o{}'x$ and $A_o^j y = A_o^j x$ for every $j \in R$.

8. Construction of solutions (y,T) of (12.1)-(12.4) by means of formula (44).

Let $R = \{1,2,\dots,r\}$ be finite and consider a point $x \in D$ such that $s = |S_x|>1$; if $s=1$, then x is D-stationary, cf. Lemma 7.6, and our construction stops.

According to § 2.7, $S=S_x$ is a subset of R such that $\{z^i : i \in R\} = \{z^i : i \in S\}$, $z^i = A^i x - b^i$, and $z^i \neq z^j$ for $i,j \in S$, $i \neq j$. It is easy to see that we may define S_x by

$$S_x = \{i \in R: \text{ there is no } t \in R \text{ such that } t<i \text{ and } z^t=z^i\}. \tag{90.1}$$

Furthermore, define J_{xi} for $i \in S_x$ by

$$J_{xi} = \{t \in R: z^t=z^i\}. \tag{90.2}$$

Obviously, it is $1 \in S_x$, $i \in J_{xi}$ for every $i \in S_x$, and $\{J_{xi}: i \in S_x\}$ is is a partition of R.

As already announced in section 6, in order to construct solutions (y,T) of (12), we consider formula (44) in Theorem 6.1 which describes the form of the matrix $T^* = (\tau_{ij})_{i \in S, j \in R}$ in an optimal solution (y^*,T^*) of the auxiliary quadratic program $(\tilde{P}^Q_{x,D})$ under the additional assumption that $\tau_{ij}>0$ for every $i \in S$ and $j \in R$.

Having a solution (y,T) of (12.1)-(12.4), then by means of (13.2), we immediately find also a solution (y,Π) of (3.1)-(3.4), see Lemma 2.4.

Formulas for $T^*=(\tau_{ij})$, which are more general than (44), may be derived from the optimality conditions for $(\tilde{P}^Q_{x,D})$ in section 5 if the condition $\tau_{ij}>0$ for all $i \in S$ and $j \in R$ is replaced by the assumption

$$\tau_{ij}=1 \text{ for } (i,j)=(i_l,j_l),\ l=1,2,\dots,l_o \tag{91.1}$$

$$\tau_{ij}=0 \text{ for } i=i_l, j\neq j_l \text{ and } i\neq i_l,\ j=j_l,\ l=1,2,\dots,l_o \tag{91.2}$$

$$\tau_{ij}>0 \text{ for } i\neq i_l,\ j\neq j_l,\ l=1,2,\ldots,l_o, \tag{91.3}$$

where (i_l,j_l), $l=1,2,\ldots,l_o$, with $0\le l_o\le s$ are given fixed elements of SxR. Now, define the index sets $S_o=S_{xo},R_o$ by

$$S_o = S_x \setminus \{i_l:\ 1\le l\le l_o\}$$
$$R_o = R \setminus \{j_l:\ 1\le l\le l_o\}, \tag{92}$$

hence $|S_o|=s-l_o$, $|R_o|=r-l_o$. Under the assumptions (91) the conditions (12.1)-(12.3) are reduced to

$$\sum_{j\in R_o} \tau_{ij} = 1,\ \tau_{ij} > 0 \text{ for all } i\in S_o,\ j\in R_o \tag{92.1}$$

$$\alpha_j = \sum_{i\in S_o} \tilde{\alpha}_i\,\tau_{ij} \text{ for all } j\in R_o \tag{92.2}$$

$$\alpha_j = \tilde{\alpha}_i \text{ for all } (i,j) = (i_l,j_l),\ l=1,\ldots,l_o \tag{92.3}$$

$$A^j y-b^j = \sum_{i\in S_o} \frac{\tilde{\alpha}_i\,\tau_{ij}}{\alpha_j} z^i \text{ for all } j\in R_o \tag{92.4}$$

$$A^j y-b^j = z^i \text{ for all } (i,j) = (i_l,j_l),\ l=1,\ldots,l_o, \tag{92.5}$$

where $\tilde{\alpha}_i = \tilde{\alpha}_i(x) = \sum_{z^t=z^i} \alpha_t = \sum_{t\in J_{xi}} \alpha_t$, cf. § 2.7.

Special cases

a) Let $l_o=s$, then it is $S_o=\emptyset$, see (92), and because of (92.2), (92.4), we must also have $R_o=\emptyset$. Hence, $l_o=s$ implies that $s=l_o=r$. Consequently, it is $S=R$, $\tilde{\alpha}_i=\alpha_i$ for each $i\in R$, and by (92.3), (92.5), we find that

$$\alpha_{j_l} = \tilde{\alpha}_{i_l} = \alpha_{i_l},\ l=1,\ldots,r \tag{92.3a}$$

$$A^{j_l}y - b^{j_l} = z^{i_l},\ l=1,\ldots,r, \tag{92.5a}$$

and therefore

$$F(y) = \sum_{j=1}^{r} \alpha_j u(A^j y-b^j) = \sum_{l=1}^{r} \alpha_{j_l} u(A^{j_l}y-b^{j_l}) =$$

$$= \sum_{l=1}^{r} \alpha_{i_l} u(z^{i_l}) = \sum_{i=1}^{r} \alpha_i u(z^i) = F(x).$$

In the non-trivial case, i.e. if $i_l \neq j_l$ for at least one $l=l_1$, $1 \leq l_1 \leq r$, then we find that $A^j y \neq A^j x$ for at least one $j \in R$, see Definition 4.1 of a stationary point x. Indeed, assuming that $A^j y = A^j x$ for all $j \in R$, then by (92.5a) we find

$$z^{j_l} = A^{j_l}x - b^{j_l} = A^{j_l}y - b^{j_l} = z^{i_l}$$

for every $l=1,\dots,r$. Since $i_l \neq j_l$ for $l=l_1$, this means that $s<r$ which is a contradiction to $s=l_o=r$.

b) Let $l_o=s-1$, then it is $|S_o|=s-l_o=1$. Hence, $S_o=\{i_o\}$ and (92.2),(92.4) yield

$$\alpha_j = \tilde{\alpha}_{i_o j}\,\tau_{i_o j} \text{ for all } j \in R_o \tag{92.2a}$$

$$A^j y - b^j = \frac{\tilde{\alpha}_{i_o j}\,\tau_{i_o j}}{\alpha_j} z^{i_o} = z^{i_o} \text{ for all } j \in R_o. \tag{92.4a}$$

Moreover, (92.1) yields

$$\sum_{j \in R_o} \tau_{i_o j} = 1,\ \tau_{i_o j} > 0 \text{ for all } j \in R_o. \tag{92.1a}$$

Consequently, as above, we find

$$F(y) = \sum_{j=1}^{r} \alpha_j u(A^j y - b^j) = \sum_{l=1}^{s-1} \alpha_{j_l} u(A^{j_l}y - b^{j_l}) + \sum_{j \in R_o} \alpha_j u(A^j y - b^j) =$$

$$= \sum_{l=1}^{s-1} \tilde{\alpha}_{i_l} u(z^{i_l}) + \sum_{j \in R_o} \tilde{\alpha}_{i_o j}\,\tau_{i_o j}\, u(z^{i_o}) =$$

$$= \sum_{l=1}^{s-1} \tilde{\alpha}_{i_l} u(z^{i_l}) + \tilde{\alpha}_{i_o j} u(z^{i_o}) = \sum_{i \in S} \tilde{\alpha}_i u(z^i) = F(x).$$

Assuming, in the present, case $l_o=s-1$ that $A^j y = A^j x$ for all $j \in R$, by (92.5) we find

$$z^{j_l} = A^{j_l}x - b^{j_l} = A^{j_l}y - b^{j_l} = z^{i_l},\ l=1,2,\dots,s-1.$$

Suppose now that there is an index $1 \leq l \leq s-1$ such that $j_l \neq i_l$. If $i_l \in R_o$, then by (92.4a) and (92.5) we obtain

$$z^{i_o} = A^{i_l}y - b^{i_l} = A^{i_l}x - b^{i_l} = z^{i_l}$$

which is a contradiction to the definition of $S=\{i_1,\dots,i_{s-1},i_o\}$. If $i_1 \notin R_o$, cf. (92), then $i_1 \in \{j_1,\dots,j_{s-1}\}$, hence, there is an index $1 \le \tilde{l} \le s-1$, $\tilde{l} \neq l$, such that $i_l = j_{\tilde{l}}$. This yields

$$z^{i_l} = z^{j_{\tilde{l}}} = z^{i_{\tilde{l}}}$$

which again contradicts the definition of S. Consequently if $i_l \neq j_l$ for at least one index $1 \le l \le s-1$, then we have $A^j y \neq A^j x$ for at least one $j \in R$, cf. Definition 4.1 of stationary points.

Assuming now that (y^*,T^*) is an optimal solution of $(\tilde{P}^Q_{x,D})$ such that $T^*=(\tau_{ij})$ fulfills (91), where $l_o, 0 \le l_o < s$, is a given integer, then by Lemma 5.2 the remaining elements τ_{ij}, $(i,j) \in S \times R$, of T^* are given, cf. section 6, in generalization of (44) by the following formulas

$$\tau_{ij} = c^o_{ij} - \frac{1}{2} u_i' v_j, \quad i \in S_o, \; j \in R_o, \tag{93.1}$$

where

$$c^o_{ij} = \frac{\alpha_j}{(s-l_o)\tilde{\alpha}_i} + \left(1 - \frac{\sum_{i \in S_o} \tilde{\alpha}_i}{(s-l_o)\tilde{\alpha}_i}\right) \frac{\alpha_j^2}{\sum_{j \in R_o} \alpha_j^2} \tag{93.2}$$

$$u_i = \frac{1}{\tilde{\alpha}_i} (z^i - \tilde{z}) \tag{93.3}$$

$$\tilde{z} = \frac{1}{s-l_o} \sum_{i \in S_o} z^i \tag{93.4}$$

$$v_j = \alpha_j \gamma_j - \alpha_j^2 \frac{\sum_{j \in R_o} \alpha_j \gamma_j}{\sum_{j \in R_o} \alpha_j^2}, \tag{93.5}$$

where γ_j, $j \in R_o$, are the Lagrange multipliers related to the corresponding equations in (12.3).

Because of (92.3), it is

$$\sum_{j \in R_o} \alpha_j = 1 - \sum_{l=1}^{l_o} \alpha_{j_l} = 1 - \sum_{l=1}^{l_o} \tilde{\alpha}_{i_l} = \sum_{i \in S_o} \tilde{\alpha}_i .$$

Hence, we find, cf. (45),

$$\sum_{j\in R_o} c^o_{ij} = 1, \quad \sum_{i\in S_o} \tilde{\alpha}_i c^o_{ij} = \alpha_j \text{ for all } i \in S_o,\ j \in R_o \tag{94.1}$$

$$\sum_{i\in S_o} \tilde{\alpha}_i u_i = 0 \tag{94.2}$$

$$\sum_{j\in R_o} v_j = 0. \tag{94.3}$$

Obviously, formulas (44) and (45) result from (93),(94), resp., by setting $l_o=0$. Corresponding to Lemma 6.1, here we obtain:

Lemma 8.1. a) If $\alpha_j = \frac{1}{r}$ for every $j \in R$, then $c^o_{ij} = \frac{1}{r-l_o}$ for all $i \in S_o$, $j \in R_o$. b) If $\min_{j\in R_o} \alpha_j + (s-l_o) \min_{i\in S_o} \tilde{\alpha}_i \geq \sum_{i\in S_o} \tilde{\alpha}_i$, then $c^o_{ij}>0$ for every $i \in S_o$, $j \in R_o$.

Proof. a) Here it is $\frac{\alpha_j^2}{\sum_{j\in R_o} \alpha_j^2} = \frac{1}{r-l_o}$, and by (92.3), we have

$\tilde{\alpha}_{i_l} = \alpha_{j_l} = \frac{1}{r}$, $l=1,\dots,l_o$. Thus, from (93.2) follows

$$c^o_{ij} = \frac{1}{r(s-l_o)\tilde{\alpha}_i} + \left(1 - \frac{\sum_{i\in S_o} \tilde{\alpha}_i}{(s-l_o)\tilde{\alpha}_i}\right) \frac{1}{r-l_o} =$$

$$= \frac{1}{r(s-l_o)\tilde{\alpha}_i} + \left(1 - \frac{1-\sum_{l=1}^{l_o} \tilde{\alpha}_{i_l}}{(s-l_o)\tilde{\alpha}_i}\right) \frac{1}{r-l_o} =$$

$$= \frac{1}{r(s-l_o)\tilde{\alpha}} + \left(1 - \frac{1-\frac{l_o}{r}}{(s-l_o)\tilde{\alpha}_i}\right) \frac{1}{r-l_o} = \frac{1}{r-l_o}$$

for every $i \in S_o$, $j \in R_o$. b) Let $i \in S_o, j \in R_o$. By (93.2), the assertion obviously holds if $(s-l_o)\tilde{\alpha}_i \geq \sum_{i\in S_o} \tilde{\alpha}_i$. Since $\frac{\alpha_j^2}{\sum_{j\in R_o} \alpha_j^2} < 1$, we find in the opposite case

$$c^o_{ij} > \frac{\alpha_j}{(s-l_o)\tilde{\alpha}_i} + 1 - \frac{\sum_{i\in S_o} \tilde{\alpha}_i}{(s-l_o)\tilde{\alpha}_i} =$$

$$= \frac{1}{(s-l_o)\tilde{\alpha}_i} (\alpha_j + (s-l_o)\tilde{\alpha}_i - \sum_{i\in S_o} \tilde{\alpha}_i)$$

$$\geq \frac{1}{(s-l_o)\tilde{\alpha}_i} (\min_{j\in R_o} \alpha_j + (s-l_o)\min_{i\in S_o} \tilde{\alpha}_i - \sum_{i\in S_o} \tilde{\alpha}_i) \geq 0.$$

Note. In the case $l_o=0$ the condition in (b) reads

$\min_{j\in R} \alpha_j + s \min_{i\in S} \tilde{\alpha}_i \geq 1.$

Our construction of solutions (y,T) of (12.1)-(12.4) is now started by inserting (93.1) into (92.4). With (93.3),(93.4) for each $j \in R_o$ we find

$$A^j y - b^j = \sum_{i\in S_o} \frac{\tilde{\alpha}_i \tau_{ij}}{\alpha_j} z^i = \sum_{i\in S_o} \frac{\tilde{\alpha}_i}{\alpha_j} (c^o_{ij} - \frac{1}{2} u_i' v_j) z^i =$$

$$= \sum_{i\in S_o} \frac{\tilde{\alpha}_i}{\alpha_j} c^o_{ij} z^i - \frac{1}{2} \sum_{i\in S_o} \frac{\tilde{\alpha}_i}{\alpha_j} z^i u_i' v_j =$$

$$= \sum_{i\in S_o} q_{ij} z^i - \frac{1}{2\alpha_j} \sum_{i\in S_o} z^i (z^i - \tilde{z})' v_j = \qquad (95.1)$$

$$= \sum_{i\in S_o} q_{ij} z^i - \frac{1}{2\alpha_j} (\sum_{i\in S_o} z^i z^{i\prime} - (s-l_o)\tilde{z}\tilde{z}') v_j =$$

$$= \sum_{i\in S_o} q_{ij} z^i - \frac{s-l_o}{2} Q_x \frac{v_j}{\alpha_j} ,$$

where q_{ij}, Q_x are defined, cf. (93.2), by

$$q_{ij} = \frac{\tilde{\alpha}_i}{\alpha_j} c^o_{ij} = \frac{1}{s-l_o} + \frac{\alpha_j}{\sum_{j\in R_o} \alpha_j^2} (\tilde{\alpha}_i - \frac{\sum_{i\in S_o} \tilde{\alpha}_i}{s-l_o}) \qquad (95.2)$$

$$Q_x = \frac{1}{s-l_o} \sum_{i\in S_o} z^i z^{i\prime} - \tilde{z}\tilde{z}' = \frac{1}{s-l_o} \sum_{i\in S_o} (z^i-\tilde{z})(z^i-\tilde{z})' \qquad (95.3)$$

Remark

a) Equations (95.2) and (94.1) yield

$$\sum_{j\in R_o} q_{ij}\, \alpha_j = \tilde{\alpha}_i,\ \sum_{i\in S_o} q_{ij} = 1 \text{ for all } i\in S_o,\ j\in R_o. \qquad (95.4)$$

b) From (92), (95.1) and (94) we obtain

$$\bar{A}y-\bar{b} = \sum_{j\in R} \alpha_j(A^j y-b^j) = \sum_{j\in R_o} \alpha_j(A^j y-b^j) + \sum_{l=1}^{l_o} \alpha_{j_l}(A^{j_l} y-b^{j_l}) =$$

$$= \sum_{j\in R_o} \alpha_j \sum_{i\in S_o} q_{ij}\, z^i - \frac{s-l_o}{2} Q_x \sum_{j\in R_o} v_j + \sum_{l=1}^{l_o} \tilde{\alpha}_{i_l} z^{i_l} =$$

$$= \sum_{i\in S_o} \tilde{\alpha}_i z^i + \sum_{l=1}^{l_o} \tilde{\alpha}_{i_l} z^{i_l} - \frac{s-l_o}{2} Q_x \sum_{j\in R_o} v_j$$

$$= \bar{A}x - \bar{b} - \frac{s-l_o}{2} Q_x \sum_{j\in R_o} v_j .$$

Hence, equation (4), i.e. $\bar{A}x = \bar{A}y$, is implied by (94.3). Conversely, (4) implies (94.3), provided that Q_x is a regular mxm matrix.

c) The symmetric nonnegative definite mxm covariance matrix Q_x is positive definite if and only if not all vectors $z^i=A^i x-b^i$ with $i\in S_o$ are contained in a certain fixed hyperplane of $\mathbb{R}^m$. If $(z^i-\tilde{z})_{i\in S_o}$ denotes the $m\times(s-l_o)$ matrix, having the columns $z^i-\tilde{z}, i\in S_o$, then the regularity of Q_x can be described by the rank condition

$$\text{rank}(z^i-\tilde{z})_{i\in S_o} = m. \qquad (96a)$$

If i_o is an arbitrary, but a fixed element of S_o, then (96a) is equivalent to

$$\operatorname{rank}((A^{i}-A^{i_o})x-(b^{i}-b^{i_o}))_{i\in S_o} = m. \tag{96b}$$

Obviously, if Q_x is positive definite, then $s-l_o>m$. Under weak assumptions, the set

$$\{x\in\mathbb{R}^n:\ Q_x \text{ is not regular}\}$$

has Lebesgue measure zero. In the important special case m=1 we have

$$Q_x = \frac{1}{s-l_o}\sum_{i\in S_o}(z^i-\tilde{z})^2,$$

and then $Q_x>0$ holds if and only if $s-l_o>1$, which simply means that $z^i \neq z^j$ for at least two elements $i,j\in S_o, i\neq j$.

Moreover, if the random matrix $(A(\omega),b(\omega))$ is given by (60), hence, if

$$(A^i,b^i) = (A^{(o)},b^{(o)}) + \sum_{t=1}^{L}\xi_t^i(A^{(t)},b^{(t)}),$$

where $(A^{(t)},b^{(t)})$, $t=0,1,\dots,L$, are given $m \times (n+1)$ matrices, and $\xi_t^i, i\in R$, are the realizations of discretely distributed random variables $\xi_t(\omega), t=1,\dots,L$, having mean zero, then Q_x can be represented by

$$Q_x = \sum_{t,\tau=1}^{L}\tilde{\operatorname{cov}}(\xi_t,\xi_\tau)(A^{(t)}x-b^{(t)})(A^{(\tau)}x-b^{(\tau)})', \tag{97}$$

where $\tilde{\operatorname{cov}}(\xi_t,\xi_\tau)$ is defined by

$$\tilde{\operatorname{cov}}(\xi_t,\xi_\tau) = \frac{1}{s-l_o}\sum_{i\in S_o}(\xi_t^i-\tilde{\xi}_t)(\xi_\tau^i-\tilde{\xi}_\tau) \tag{97.1}$$

with

$$\tilde{\xi}_t = \frac{1}{s-l_o}\sum_{i\in S_o}\xi_t^i,\ t,\tau=1,\dots,L. \tag{97.2}$$

In the special case

$$\tilde{\operatorname{cov}}(\xi_t,\xi_\tau) = V_t\,\delta_{t\tau}, V_t>0, t,\tau=1,\dots,L, \tag{97.3}$$

Q_x has the simple form

$$Q_x = \sum_{t=1}^{L} v_t(A^{(t)}x-b^{(t)})(A^{(t)}x-b^{(t)})'.$$

In this case, Q_x is regular if and only if not all vectors $A^{(t)}x-b^{(t)}, t=1,\dots,L$, are contained in a certain fixed hyperplane, which can be characterized by the rank condition

$$\text{rank}(A^{(1)}x-b^{(1)},\dots,A^{(L)}x-b^{(L)}) = m,$$

cf. (96). E.g. (97.3) holds if $l_o=0, S=R, \alpha_j=\frac{1}{r}$ for all $j \in R$ and $\xi_1,\xi_2,\dots,\xi_L$ are uncorrelated random variables.

Because of condition (4), i.e. $\bar{A}y=\bar{A}x$, in the following we suppose that the rank condition

$$\bar{m} = \text{rank } \bar{A} < n \tag{98}$$

holds. Since, in practice, it is very often $m<n$ or even $m<<n$, (98) is only a very weak assumption.

If Q_x is positive definite, then $s-l_o \geq 2$ and (95.1) yields

$$v_j = \frac{2\alpha_j}{s-l_o} Q_x^{-1}\left(\sum_{i\in S_o} q_{ij}\, z^i + b^j - A^j y\right),\ j \in R_o, \tag{99}$$

and from (93.1) and (93.3) we then obtain

$$\tau_{ij} = c_{ij}^o - \frac{\alpha_j}{(s-l_o)\tilde{\alpha}_i} (z^i-\tilde{z})Q_x^{-1}\left(\sum_{i\in S_o} q_{ij}\, z^i+b^j-A^j y\right) \tag{100}$$

for all $i \in S_o, j \in R_o$. Therefore, if Q_x is regular, then the condition $\tau_{ij} \geq (>) 0$ holds for an index pair $(i,j) \in S_o \times R_o$ if and only if y satisfies the linear inequality

$$(z^i-\tilde{z})'Q_x^{-1}\left(\sum_{i\in S_o} q_{ij}\, z^i+b^j-A^j y\right) \leq (<) \frac{(s-l_o)\tilde{\alpha}_i}{\alpha_j} c_{ij}^o. \tag{101}$$

Now, we can formulate this first result:

<u>Theorem 8.1.</u> Consider a point $x \in \mathbb{R}^n$ such that Q_x is positive definite. Furthermore, let $(i_l,j_l) \in S\times R$, $l=1,\dots,l_o$, $0\leq l_o<s$, be an empty ($l_o=0$) or nonempty ($0<l_o<s$) set of index pairs such that $i_l \neq i_\lambda, l\neq\lambda$ and $j_l \neq j_\lambda$ for $l\neq\lambda$. For $l_o>0$ assume that

$$\tilde{\alpha}_{i_l} = \alpha_{j_l}, \; l=1,\ldots,l_o. \tag{102.1}$$

a) If $y \in \mathbb{R}^n$ satisfies the system of linear equations (4) and (92.5), i.e.

$$\bar{A}y = \bar{A}x \tag{102.2}$$

$$A^{j_l}y-b^{j_l} = A^{i_l}x-b^{i_l}, \; l=1,\ldots,l_o, \tag{102.3}$$

as well as the linear inequality (101)

$$(z^i-\tilde{z})'Q_x^{-1}\Big(\sum_{i \in S_o} q_{ij}\; z^i+b^j-A^jy\Big)\leq(<)\; \frac{(s-l_o)\tilde{\alpha}_i}{\alpha_j}\, c_{ij}^o \tag{102.4}$$

for every $(i,j) \in S_o \times R_o$, then (y,T) fulfills (12.1)-(12.3), where $T = (\tau_{ij})$ is given by (91.1),(91.2) and (100).

b) If, in addition to the above conditions (102), the inequality (102.4) holds with the strict inequality sign for at least two pairs $(i_1,j),(i_2,j) \in S_o \times R_o$, $i_1 \neq i_2$, then (y,T) fulfills also (12.1)-(12.4a). If $l_o=0$ and, if in addition to the above conditions (102), there is an index $j \in R_o=R$ such that (102.4) holds with the strict inequality sign for every (i,j) with $i \in S_o=S$, then (y,T) fulfills (12.1)-(12.4b).

Proof. From the assumptions in (102), we obtain that $\tau_{ij} \geq 0$ for all $i \in S, j \in R$ and $\sum_{j \in R} \tau_{i_l j}=1$,

$$\sum_{i \in S} \tilde{\alpha}_i \tau_{ij_l} = \tilde{\alpha}_{i_l}\tau_{i_l j_l} = \tilde{\alpha}_{i_l} = \alpha_{j_l}, \quad \sum_{i \in S} \frac{\tilde{\alpha}_i \tau_{ij_l}}{\alpha_{j_l}} z^i = \frac{\tilde{\alpha}_{i_l} \tau_{i_l j_l}}{\alpha_{j_l}} z^{i_l} =$$

$z^{i_l} = A^{j_l}y-b^{j_l}$ for each $l=1,\ldots,l_o$. If $i \in S_o$, then, by (100) and (91.2),(94.1),(95.4), we find

$$\sum_{j\in R} \tau_{ij} = \sum_{j\in R_o} \tau_{ij} = \sum_{j\in R_o} c^o_{ij}$$

$$- \frac{1}{(s-l_o)\tilde{\alpha}_i} (z^i-\tilde{z})'Q_x^{-1}(\sum_{i\in S_o} (\sum_{j\in R_o} q_{ij} \alpha_j)z^i + \sum_{j\in R_o} \alpha_j(b^j-A^jy)) =$$

$$= 1 - \frac{1}{(s-l_o)\tilde{\alpha}_i}(z^i-\tilde{z})'Q_x^{-1}(\sum_{i\in S_o} \tilde{\alpha}_i z^i - \sum_{j\in R_o} \alpha_j(A^jy-b^j)) =$$

$$= 1 - \frac{1}{(s-l_o)\tilde{\alpha}_i} (z^i-\tilde{z})'Q_x^{-1}(\bar{A}x-\bar{b}- \sum_{l=1}^{l_o} \tilde{\alpha}_{i_l} z^{i_l} -(\bar{A}y-\bar{b}- \sum_{l=1}^{l_o} \alpha_{j_l}(A^{j_l}y-b^{j_l}))) =$$

$= 1.$

For $j \in R_o$, using (100),(91.2),(93.4) and (94.1), it is

$$\sum_{i\in S} \tilde{\alpha}_i \tau_{ij} = \sum_{i\in S_o} \tilde{\alpha}_i \tau_{ij} =$$

$$= \sum_{i\in S_o} \tilde{\alpha}_i c^o_{ij} - \frac{\alpha_j}{s-l_o} \sum_{i\in S_o} (z^i-\tilde{z})'Q_x^{-1}(\sum_{i\in S_o} q_{ij} z^i+b^j-A^jy) =$$

$$= \alpha_j.$$

Furthermore, if $j \in R_o$, then (100) and (91.2),(95.2),(95.3) yield

$$\sum_{i\in S} \frac{\tilde{\alpha}_i\tau_{ij}}{\alpha_j} z^i = \sum_{i\in S_o} \frac{\tilde{\alpha}_i \tau_{ij}}{\alpha_j} z^i =$$

$$= \sum_{i\in S_o} \frac{\tilde{\alpha}_i}{\alpha_j}(c^o_{ij} - \frac{\alpha_j}{(s-l_o)\tilde{\alpha}_i}(z^i-\tilde{z})'Q_x^{-1}(\sum_{i\in S_o} q_{ij} z^i+b^j-A^jy))z^i =$$

$$= \sum_{i\in S_o} q_{ij} z^i - \frac{1}{s-l_o} \sum_{i\in S_o} z^i(z^i-\tilde{z})'Q_x^{-1}(\sum_{i\in S_o} q_{ij} z^i-b^j-A^jy) =$$

$$= \sum_{i\in S_o} q_{ij} z^i - (\frac{1}{s-l_o} \sum_{i\in S_o} z^iz^{i'}-\tilde{z}\tilde{z}')Q_x^{-1}(\sum_{i\in S_o} q_{ij} z^i+b^j-A^jy) =$$

$$= \sum_{i\in S_o} q_{ij} z^i - Q_x Q_x^{-1}(\sum_{i\in S_o} q_{ij} z^i + b^j - A^j y) = A^j y - b^j,$$

which now shows that (y,T) satisfies (12.1)-(12.3). If the additional assumptions in the second part of the theorem hold true, then either $\tau_{i_1 j}>0$, $\tau_{i_2 j}>0$ for at least two pairs (i_1,j), $(i_2,j)\in S_o \times R_o$, $i_1 \neq i_2$, or there exists an index $j\in R$ such that $\tau_{ij}>0$ for all $i\in S$. Consequently, (y,T) fulfills in this case (12.1)-(12.4a),(12.1)-(12.4b), respectively.

Remark

a) If there is a feasible domain $D \neq \mathbb{R}^n$, then we have also to fulfill (12.5), i.e. to (102.1)-(102.4) we must add the constraint

$$y\in D. \tag{102.5}$$

b) For $l_o=0$ it is $S_o=S, R_o=R$ and the conditions (102.1), (102.3) are cancelled out.

c) The relations (102.1)-(102.5) can be also interpreted system of

- conditions for the vector x, where y is then a *given* vector, or
- conditions for the tuple (x,y).

According to Theorem 8.1, the main problem is the solution of inequality (102.4) for y with given x, for x with given y, or for the tuple (x,y), respectively. Supposing that (102.1)-(102.3) hold true, we first note that

$$\sum_{i\in S_o} (z^i-\tilde{z}) = \sum_{i\in S_o} z^i-(s-l_o)\tilde{z} = 0.$$

Defining $\tilde{A},\tilde{b}$ by

$$\begin{aligned} \tilde{A} &= \tilde{A}_{S_o} = \frac{1}{s-l_o} \sum_{i\in S_o} A^i \\ \tilde{b} &= \tilde{b}_{S_o} = \frac{1}{s-l_o} \sum_{i\in S_o} b^i, \end{aligned} \tag{103}$$

we have

$$z^i - \tilde{z} = (A^i - \tilde{A})x - (b^i - \tilde{b}).$$

From (95.2) and with (103) we get

$$\sum_{i \in S_o} q_{ij} z^i = \sum_{i \in S_o} \left(\frac{1}{s-l_o} + \frac{\alpha_j}{\sum_{j \in R_o} \alpha_j^2}\left(\tilde{\alpha}_i - \frac{\sum_{i \in S_o} \tilde{\alpha}_i}{s-l_o}\right)\right) z^i =$$

$$= \tilde{A}x - \tilde{b} + \hat{\alpha}_j\left(\sum_{i \in S_o} \tilde{\alpha}_i z^i - \left(\sum_{i \in S_o} \tilde{\alpha}_i\right)(\tilde{A}x - \tilde{b})\right), \qquad (104.1)$$

$$= \left(1 - \hat{\alpha}_j \sum_{i \in S_o} \tilde{\alpha}_i\right)(\tilde{A}x - \tilde{b}) + \hat{\alpha}_j \sum_{i \in S_o} \tilde{\alpha}_i z^i,$$

where $\hat{\alpha}_j$ is given by

$$\hat{\alpha}_j = \frac{\alpha_j}{\sum_{j \in R_o} \alpha_j{}^2}, \quad j \in R_o. \qquad (104.2)$$

Hence, inequality (102.4) has the form

$$(z^i - \tilde{z})' Q_x^{-1}\left(\left(1 - \hat{\alpha}_j \sum_{i \in S_o} \tilde{\alpha}_i\right)(\tilde{A}x - \tilde{b}) + \hat{\alpha}_j \sum_{i \in S_o} \tilde{\alpha}_i z^i + \right. \qquad (102.4a)$$

$$\left. + b^j - A^j y\right) \leq (<) \frac{(s-l_o)\tilde{\alpha}_i}{\alpha_j} c_{ij}^o.$$

Under the conditions (102.1)-(102.3) it is

$$\sum_{i \in S_o} \tilde{\alpha}_i z^i = \bar{A}x - \bar{b} - \sum_{l=1}^{l_o} \tilde{\alpha}_{i_l} z^{i_l} =$$

$$= \bar{A}y - \bar{b} - \sum_{l=1}^{l_o} \alpha_{j_l}(A^{j_l} y - b^{j_l}) = \sum_{j \in R_o} \alpha_j (A^j y - b^j). \qquad (104.3)$$

In the case $l_o = 0$ we have $S_o = S, R_o = R$ and, therefore, by (104.1)

$$\sum_{i \in S} q_{ij} z^i = (1 - \hat{\alpha}_j)(\tilde{A}x - \tilde{b}) + \hat{\alpha}_j(\bar{A}x - \bar{b}). \qquad (104.4)$$

If $\alpha_j = \frac{1}{r}$ for all $j \in R$, then $\hat{\alpha}_j = \frac{r}{r-l_o}$ for all $j \in R_o$ as well as

$\tilde{\alpha}_{i_l} = \alpha_{j_l} = \frac{1}{r}$, $l=1,\ldots,l_o$. Since $\sum\limits_{i\in S_o} \tilde{\alpha}_i = 1 - \sum\limits_{l=1}^{l_o} \tilde{\alpha}_{i_l} = 1 - \frac{l_o}{r}$,

we find $1-\hat{\alpha}_j \sum\limits_{i\in S_o} \tilde{\alpha}_i = 0$ and, therefore, by (104.1),(104.3)

$$\sum_{i\in S_o} q_{ij}\, z^i = \frac{r}{r-l_o} \sum_{i\in S_o} \tilde{\alpha}_i\, z^i = \frac{1}{r-l_o} \sum_{j\in R_o} (A^j y - b^j). \qquad (104.5)$$

Finally, if $\alpha_j = \frac{1}{r}$ for all $j \in R$ and $l_o=0$, then

$$\sum_{i\in S} q_{ij}\, z^i = \bar{A}x - \bar{b} = \bar{A}y - \bar{b}. \qquad (104.6)$$

In the case $c^o_{ij} \geq 0$, cf. Lemma 8.1, inequality (102.4) is implied by the norm condition

$$||z^i - \tilde{z}|| \cdot ||Q_x^{-1}|| \cdot || \sum_{i\in S_o} q_{ij}\, z^i + b^j - A^j y|| \leq (<) \frac{(s-l_o)\tilde{\alpha}_i}{\alpha_j}\, c^o_{ij} \qquad (105)$$

Note, that according to (100), τ_{ij} can be estimated from below by

$$\tau_{ij} \geq c^o_{ij} - \frac{\alpha_j}{(s-l_o)\tilde{\alpha}_i} ||z^i - \tilde{z}|| \cdot ||Q_x^{-1}|| \cdot || \sum_{i\in S_o} q_{ij}\, z^i + b^j - A^j y||. \qquad (106)$$

Consider now a family p of parameters

$$p = (S,\ (J_i)_{i\in S}, \bar{a}, l_o, (i_l, j_l), a^{i_l}, l=1,\ldots,l_o) \qquad (107)$$

composed of

- a set $S \subset R$ with $1 \in S$ and $s=|S|>1$
- a partition $J_i, i \in S$, of R with $i \in J_i$ for all $i \in S$
- an m-vector $\bar{a}$
- an integer $l_o, 0 \leq l_o < s-1$

and for $l_o > 0$

- a set of index pairs $(i_l, j_l) \in S \times R, l=1,\ldots,l_o$, such that $i_l \neq i_\lambda, j_l \neq j_\lambda$ for $l \neq \lambda$ and $\sum\limits_{t\in J_{i_l}} \alpha_t = \alpha_{i_l}$ for all $l=1,\ldots,l_o$
- m-vectors $a^{i_l}, l=1,\ldots,l_o$.

Given a family p of parameters according to (107), then, by X(p) we denote the set of n-vectors x

$$X(p) = \{x \in \mathbb{R}^n : S_x = S,\ J_{xi} = J_i \text{ for all } i \in S,$$
$$\bar{A}x = \bar{a},\ A^{i_l}x = a^{i_l},\ l=1,\dots,l_o, \tag{108}$$
$$Q_x \text{ is regular}\}.$$

Remark

a) If $l_o=0$, then the equations $A^{i_l}x=a^{i_l}, l=1,\dots,l_o$, are cancelled. b) If $\bar{A}=0$, then the vector $\bar{a}$ in p and the constraint $\bar{A}x = \bar{a}$ in (108) are cancelled. c) Obviously, for every n-vector x, there is a parameter family p such that the first four conditions in (108) hold. d) Q_x is regular if and only if (96) holds. Further sufficient conditions for the regularity of Q_x are given later on.

Now, we want discuss the first two conditionsin (108) in more detail. For a set $S \subset R$ and a partition J_i, $i \in S$, of R according to (107), define

$$H(S,(J_i)_{i \in S}) = \{x \in \mathbb{R}^n : S_x=S,\ J_{xi}=J_i \text{ for each } i \in S\}. \tag{109}$$

According to (90), a vector x is an element of this set if and only if $z^i \neq z^j$ for all $i,j \in S$, $i \neq j$, and $z^j=z^i$ for every $j \in J_i$, $i \in S$. Thus, it is

$$H(S,(J_i)_{i \in S}) = \{x \in \mathbb{R}^n : x \notin H^{ij} \text{ for all } i,j \in S, i \neq j, \tag{109a}$$
$$x \in H^{ij} \text{ for all } j \in J_i, i \in S\},$$

where $H^{ij} = H^{ji}$ denote the linear manifolds

$$H^{ij} = \{x \in \mathbb{R}^n : z^i=z^j\} = \{x \in \mathbb{R}^n : (A^i-A^j)x = (b^i-b^j)\} \tag{110}$$

in $\mathbb{R}^n$. For $i \neq j$ it is $(A^i,b^i) \neq (A^j,b^j)$, hence, $H^{ij} = \emptyset$ or H^{ij} is a proper linear submanifold of $\mathbb{R}^n$.

Consequently, for every subset $S \subset R, S \neq R$, and every partition $(J_i)_{i \in S}$ of R with $i \in J_i$ for $i \in S$ it is

$$H(S,(J_i)_{i\in S}) \subset \bigcap_{\substack{j\in J_i \\ i\in S}} H^{ij}, \tag{111.1}$$

hence, $H(S,(J_i)_{i\in S})$ is then contained in an intersection of lower dimensional linear submanifolds of $\mathbb{R}^n$. Furthermore, if $S=R$ and, therefore, $J_i=\{i\}$ for each $i \in S$, then

$$H(R,(\{i\})_{i\in R}) = \mathbb{R}^n \setminus \bigcup_{i \neq j} H^{ij}. \tag{111.2}$$

In the important special case that $b^j=b_o$ for each $j \in R$, where b_o is a given fixed m-vector, we have

$\lambda x \in H(S,(J_i)_{i\in S})$ for each $x \in H(S,(J_i)_{i\in S})$ and $\lambda \in \mathbb{R}$, $\lambda \neq 0$,

where this then holds for every $S \subset R$ and every partition $(J_i)_{i\in S}$.

Finally, we note that $x \in H(S_x,(J_{xi})_{i\in S_x})$ for each $x \in \mathbb{R}^n$, and there is only a finite number of sets of the type $H(S,(J_i)_{i\in S})$.

Given a parameter family p, suppose now, that for a vector $x \in X(p)$, cf.(107),(108), there exists a vector $y \in \mathbb{R}^n$ satisfying the following conditions

$$\bar{A}y = \bar{a} \tag{112.1}$$

$$A^{j_l}y-b^{j_l} = a^{i_l}-b^{i_l}, \; l=1,\ldots,l_o \tag{112.2}$$

$$(z^i-\tilde{z})'Q_x^{-1}\Big(\sum_{i\in S_o} q_{ij}\, z^i+b^j-A^jy\Big) < \frac{(s-l_o)\tilde{\alpha}_i}{\alpha_j}\, c^o_{ij}, i \in S_o, j \in R_o, \tag{112.3a}$$

$$||z^i-\tilde{z}||\cdot||Q_x^{-1}||\cdot\Big|\Big|\sum_{i\in S_o} q_{ij}\, z^i+b^j-A^jy\Big|\Big| < \frac{(s-l_o)\tilde{\alpha}_i}{\alpha_j}\, c^o_{ij}, \; i \in S_o, j \in R_o, \tag{112.3b}$$

respectively; in (112.3b), which implies (112.3a), we always suppose that $c^o_{ij} > 0$ for all $i \in S_o, j \in R_o$. Hence, if $T = (\tau_{ij})$ is given by (91.1), (91.2) and (100), then, according to Theorem 8.1, we know that (y,T) fulfills (12.1)-(12.4a), where $|S_o| \geq 2$ and

$\tau_{ij}>0$ for all $i\in S_o, j\in R_o$. Since $T \neq T^o$, cf. (14), by Corollary 4.1, there is at least one index $j\in R$ such that $A^j y \neq A^j x$, hence, $y\neq x$. If $h=y-x$ is a feasible direction for D at x, e.g. if $x\in \overset{o}{D}$ (= interior of D), then by Definition 4.1 or Theorem 7.4 this means that x is not D-stationary.

Hence, we arrive at this theorem:

Theorem 8.2. Suppose that $x\in X(p)$ for some parameter p, see (107), (108). If y satisfies (112), then there is a matrix $T \neq T^o$ such that (y,T) satisfies (12.1)-(12.4a), where $A^j y \neq A^j x$ for at least one $j\in R$. Furthermore, if, for $x\in D\cap X(p)$, (112) has a solution y such that $y-x$ is a feasible direction for D at x, then x is not D-stationary.

The above theorem yields a numerical stationarity criterion.

8.1. A numerical stationarity criterion. Based on the system of relations (112), on X(p) we now define functions $I_1(x)$, $I_2(x)$ and $J_1(x)$, $J_2(x)$ as follows:

$$I_1(x) = \min_{\substack{\bar{A}y = \bar{a} \\ A^{j_l}y-b^{j_l}=a^{i_l}-b^{i_l}, \\ l=1,2,\dots,l_o}} \max_{\substack{i\in S \\ j\in R_o^o}} (z^i-\tilde{z})'Q_x^{-1}\Big(\sum_{i\in S_o} q_{ij}\, z^i+b^j-A^j y\Big), \tag{113.1}$$

$$I_2(x) = \min_{\substack{i\in S_o \\ j\in R_o}} \frac{(s-l_o)\tilde{\alpha}_i}{\alpha_j}\, c_{ij}^o, \tag{113.2}$$

and

$$J_1(x) = \min_{\substack{\bar{A}y=\bar{a} \\ A^{j_l}y-b^{j_l}=a^{i_l}-b^{i_l}, \\ l=1,2,\dots,l_o}} \max_{j\in R_o} \alpha_j \Big\|\sum_{i\in S_o} q_{ij}\, z^i+b^j-A^j y\Big\|, \tag{113.3}$$

$$J_2(x) = \frac{1}{||Q_x^{-1}||} \min_{\substack{i\in S_o \\ j\in R_o^o}} \frac{(s-l_o)\tilde{\alpha}_i c_{ij}^o}{||z^i-\tilde{z}||} =$$

$$= \frac{1}{||Q_x^{-1}||} \cdot \frac{1}{\max\limits_{\substack{i\in S_o \\ j\in R_o^o}} \frac{||z^i-\tilde{z}||}{(s-l_o)\tilde{\alpha}_i c_{ij}^o}}, \tag{113.4}$$

where we suppose that the minimum in (113.1),(113.2) is attained at a certain point $\hat{y} = \hat{y}(I_1(x)),\hat{y}(J_1(x))$, respectively.

If a vector $x \in X(p)$ satisfies the inequality $I_1(x) < I_2(x)$, then there is a vector $\hat{y}$ such that $\bar{A}\hat{y}=\bar{a}$, $A^{j_l}\hat{y}-b^{j_l}=a^{i_l}-b^{i_l}$, $l=1,2,\ldots,l_o$, and

$$(z^i-\tilde{z})'Q_x^{-1}(\sum_{i\in S_o} q_{ij} z^i+b^j-A^j\hat{y}) \leq \max_{\substack{i\in S_o \\ j\in R_o^o}} (z^i-\tilde{z})'Q_x^{-1}(\sum_{i\in S_o} q_{ij} z^i+b^j-A^j\hat{y}) =$$

$$= I_1(x) < I_2(x) \leq \frac{(s-l_o)\tilde{\alpha}_i}{\alpha_j} c_{ij}^o$$

for all $i \in S_o, j \in R_o$. Hence, $\hat{y}$ fulfills (112), and by Theorem 8.2, we know then that $(\hat{y},T)$ fulfills (12.1)-(12.4a), where $T = (\tau_{ij})$ is given by (91.1),(91.2) and (100). Moreover, we find that $A^j\hat{y} \neq A^jx$ for at least one $j \in R$. The same conclusion is obtained if $x \in X(p)$ is a vector such that $J_1(x)<J_2(x)$. Hence, we reach the following corollary:

<u>Corollary 8.1.</u> Let $x \in \overset{o}{D} \cap X(p)$ for some parameter family p. a) If $I_1(x)=0$ or $J_1(x)=0$, then x is not D-stationary. b) If x is D-stationary, then

$$I_1(x) \geq I_2(x) \tag{114.1}$$

and

$$J_1(x) \geq J_2(x). \tag{114.2}$$

Remark

a) Obviously, (144.1) and (114.2) are numerical stationarity criteria! b) $I_1(x)$, $J_1(x)$, resp., is the optimal value of a linear, convex program, respectively. c) Instead of $J_1(x)$ we may also use the function

$$J_1^q(x) = \min_{\substack{\bar{A}y=\bar{a} \\ A^{j_l}y-b^{j_l}=a^{i_l}-b^{i_l}, \\ l=1,2,\ldots,l_o}} \sqrt{\sum_{j\in R_o} \alpha_j^2 \| \sum_{i\in S_o} q_{ij} \, z^i+b^j-A^jy\|^2} \tag{113.3a}$$

which is the optimal value of a quadratic program. Furthermore, it is $J_1(x) \leq J_1^q(x) \leq \sqrt{r-l_o}\, J_1(x)$. d) The constraints $A^{j_l}y-b^{j_l}=a^{i_l}-b^{i_l}$, $l=1,\ldots,l_o$, the constraint $\bar{A}y=\bar{a}$, resp., in (113) are cancelled if $l_o=0$, $\bar{A}=0$, respectively.

Before we continue the general discussion of (102.4) and (105) we first consider some important special cases.

8.2. Realizations (A^j,b^j) of $(A(\omega),b(\omega))$ with equal probabilities. Let be $\alpha_j = \frac{1}{r}$, $j=1,2,\ldots,r$, and suppose that (102.1)-(102.3) hold true. In this case, according to (104.2) and Lemma 8.1, we have that $\hat{\alpha}_j = \frac{r}{r-l_o}$ for all $j \in R_o$ and $c_{ij}^o = \frac{1}{r-l_o}$ for all $(i,j) \in R_o \times S_o$. Furthermore, under conditions (102.1) - (102.3), from (104.1) - (104.3) we obtain

$$\sum_{i\in S_o} q_{ij}\, z^i = \begin{cases} \frac{r}{r-l_o} \sum_{i\in S_o} \tilde{\alpha}_i \, z^i = \frac{1}{r-l_o} \sum_{j\in R_o} (A^jy-b^j), \text{ if } l_o \geq 0 \\ \\ \bar{A}x-\bar{b} = \bar{A}y-\bar{b}, \text{ if } l_o=0 \end{cases} \tag{115}$$

According to (93.4), (103) and (115), inequality (102.4) has the form

$$(A^i x-b^i-(\tilde{A}x-\tilde{b}))'Q_x^{-1}\left(\frac{r}{r-l_o}\sum_{i\in S_o}\tilde{\alpha}_i\; z^i+b^j-A^j y\right)\leq(<)(s-l_o)\tilde{\alpha}_i\,\frac{r}{r-l_o} \tag{116.1}$$

or

$$((A^i-\tilde{A})x-(b^i-\tilde{b}))'Q_x^{-1}\left(\left(\left(\frac{1}{r-l_o}\sum_{j\in R_o}A^j\right)-A^j\right)y-\right. \tag{116.2}$$

$$\left.\left(\left(\frac{1}{r-l_o}\sum_{j\in R_o}b^j\right)-b^j\right)\right)\leq(<)(s-l_o)\tilde{\alpha}_i\,\frac{r}{r-l_o}\;.$$

From Theorem 8.1 we now obtain this corollary:

<u>Corollary 8.2.</u> Let $\alpha_j=\frac{1}{r}$ for all $j\in R$. Consider a point $x\in\mathbb{R}^n$ such that Q_x is positive definite. Furthermore, let $(i_l,j_l)\in S\times R$, $l=1,\ldots,l_o$, with $0\leq l_o<s$ be an empty ($l_o=0$) or nonempty ($0<l_o<s$) set of index pairs. For $l_o>0$ assume that (102.1) holds. a) If $y\in\mathbb{R}^n$ satisfies the relations (102.2), (102.3) and (116) for every $(i,j)\in S_o\times R_o$, then (y,T) is a solution of (12.1)-(12.3), where T is given by (91.1),(91.2) and (100). b) If, in addition to the conditions in (a), (116) holds true with the strict inequality sign for at least two pairs $(i_1,j),(i_2,j)\in S_o\times R_o$, $i_1\neq i_2$, then (y,T) fulfills (12.1)-(12.4a). If $l_o=0$, and there is an index $j\in R_o=R$ such that, in addition to the assumptions in (a), (116) holds with the strict inequality sign for every (i,j) with $i\in S_o=S$, then (y,T) fulfills (12.1) - (12.4b).

If $\alpha_j=\frac{1}{r}$ and $b_j=b_o$ for all $j\in R$, where b_o is a fixed m-vector, e.g. $b_o=0$, then (116) is reduced to

$$(A^i x-\tilde{A}x)'Q_x^{-1}\left(\frac{r}{r-l_o}\sum_{i\in S_o}\tilde{\alpha}_i\;A^i x-A^j y\right)\leq(<)(s-l_o)\tilde{\alpha}_i\,\frac{r}{r-l_o} \tag{117.1}$$

or

$$(A^i x-\tilde{A}x)'Q_x^{-1}\left(\left(\frac{1}{r-l_o}\sum_{j\in R_o}A^j\right)-A^j\right)y\leq(<)(s-l_o)\tilde{\alpha}_i\,\frac{r}{r-l_o}\;. \tag{117.2}$$

It is easy to see that inequality (117.2) is implied, cf. (105), by the following inequality

$$||((\frac{1}{r-l_o} \sum_{j\in R_o} A^j)-A^j)'Q_x^{-1}(A^i x-\tilde{A}x)||\cdot||y|| \leq (<)(s-l_o)\tilde{\alpha}_i \frac{r}{r-l_o} \tag{118}$$

for $||y||$.

Assuming $\alpha_j = \frac{1}{r}$, $b^j=b_o$ for all $j \in R$ and $\sum_{i\in S_o} \tilde{\alpha}_i A^i = \frac{1}{r} \sum_{j\in R_o} A^j = 0$,

which holds e.g. if $l_o=0$ and $\bar{A}=0$, inequality (117) is reduced to

$$(\tilde{A}x-A^i x)'Q_x^{-1} A^j y \leq (<)(s-l_o)\tilde{\alpha}_i \frac{r}{r-l_o} \tag{119}$$

which is implied by

$$||A^{j\,\prime} Q_x^{-1}(\tilde{A}x-A^i x)||\cdot||y|| \leq (<)(s-l_o)\tilde{\alpha}_i \frac{r}{r-l_o}\ . \tag{119.1}$$

If $\alpha_j = \frac{1}{r}$, $j=1,2,\dots,r$, and $S = R$, then $\tilde{\alpha}_i = \alpha_i = \frac{1}{r}$ for all $i \in S = R$ and (116) has the form

$$((A^i-\tilde{A})x-(b^i-\tilde{b}))'Q_x^{-1}((\tilde{A}-A^j)y-(\tilde{b}-b^j)) \leq (<)1, \tag{120}$$

where $\tilde{A},\tilde{b}$ are given by (103). If $l_o=0$, then in the present case it is $\tilde{A}=\bar{A},\tilde{b}=\bar{b}$.

8.3. The case m = 1

8.3.1. Solving (102.1) - (102.4) for y with given x. Next to, let x be a given vector such that $|S_o|=s-l_o>1$, where S_o is again defined by (92). Assuming that (102.1) holds, we consider then the vectors y fulfilling (102.2),(102.3). We find $Q_x>0$, and (102.4) reads

$$(z^i-\tilde{z})(\sum_{i\in S_o} q_{ij}\ z^i+b^j-A^j y) \leq (<)\frac{(s-l_o)\tilde{\alpha}_i}{\alpha_j}\ c_{ij}^o\ Q_x. \tag{121}$$

In the case $c_{ij}^o>0$ for all $(i,j) \in S_o \times R_o$, cf. Lemma 8.1., inequality (121) reads

$$\sum_{i\in S_o} q_{ij}\, z^i + b^j - A^j y \begin{cases} \leq(<) \dfrac{(s-l_o)\tilde{\alpha}_i}{\alpha_j}\, c^o_{ij}\, \dfrac{Q_x}{z^i-\tilde{z}}, & \text{if } z^i-\tilde{z}>0 \\ \\ \geq(>) \dfrac{(s-l_o)\tilde{\alpha}_i}{\alpha_j}\, c^o_{ij}\, \dfrac{Q_x}{z^i-\tilde{z}}, & \text{if } z^i-\tilde{z}<0, \end{cases} \tag{121a}$$

where there is no constraint for $z^i-\tilde{z}=0$. Since $|S_o|>1$ and $\sum_{i\in S_o}(z^i-\tilde{z})=0$, the set of integers such that $z^i-\tilde{z}>0$, $z^i-\tilde{z}<0$, resp., is nonempty. By (100), the inequalities $\tau_{ij}\geq(>)0$ hold for all $(i,j)\in S_o \times R_o$ if (121) holds for all $(i,j)\in S_o \times R_o$. The latter is true if and only if

$$\frac{s-l_o}{\alpha_j} Q_x \max_{\substack{i\in S_o \\ z^i<\tilde{z}}} \frac{\tilde{\alpha}_i\, c^o_{ij}}{z^i-\tilde{z}} \leq(<) \sum_{i\in S_o} q_{ij}\, z^i + b^j - A^j y \leq(<) \frac{s-l_o}{\alpha_j} Q_x \min_{\substack{i\in S_o \\ z^i>\tilde{z}}} \frac{\tilde{\alpha}_i\, c^o_{ij}}{z^i-\tilde{z}} \tag{122}$$

holds for every $j\in R_o$. If $c^o_{ij}>0$ for all $(i,j)\in S_o \times R_o$, then

$$\min_{\substack{i\in S_o \\ z^i>\tilde{z}}} \frac{\tilde{\alpha}_i\, c^o_{ij}}{z^i-\tilde{z}} = \frac{1}{\max\limits_{\substack{i\in S_o \\ z^i>\tilde{z}}} \dfrac{z^i-\tilde{z}}{\tilde{\alpha}_i\, c^o_{ij}}} = \frac{1}{\max\limits_{i\in S_o} \dfrac{z^i-\tilde{z}}{\tilde{\alpha}_i\, c^o_{ij}}}$$

as well as

$$\max_{\substack{i\in S_o \\ z^i<\tilde{z}}} \frac{\tilde{\alpha}_i\, c^o_{ij}}{z^i-\tilde{z}} = \frac{1}{\min\limits_{\substack{i\in S_o \\ z^i<\tilde{z}}} \dfrac{z^i-\tilde{z}}{\tilde{\alpha}_i\, c^o_{ij}}} = \frac{1}{\min\limits_{i\in S_o} \dfrac{z^i-\tilde{z}}{\tilde{\alpha}_i\, c^o_{ij}}}\ .$$

Hence, if $c^o_{ij}>0$ for all $(i,j)\in S_o \times R_o$, then (122) reads

$$\frac{s-l_o}{\alpha_j} \frac{Q_x}{\min\limits_{i\in S_o} \frac{z^i-\tilde z}{\tilde\alpha_i\ c^o_{ij}}} \leq (<) \sum_{i\in S_o} q_{ij}\ z^i+b^j-A^j y \tag{123}$$

$$\leq (<) \frac{s-l_o}{\alpha_j} \frac{Q_x}{\max\limits_{i\in S_o} \frac{z^i-\tilde z}{\tilde\alpha_i\ c^o_{ij}}} .$$

Note, that under conditions (102.1)-(102.3) according to (104.1) and (104.3) we find that

$$\sum_{i\in S_o} q_{ij}\ z^i = (1-\hat\alpha_j \sum_{i\in S_o} \tilde\alpha_i)(\tilde A x-\tilde b)+\hat\alpha_j \sum_{j\in R_o} \alpha_j(A^j y-b^j). \tag{123.1}$$

It is easy to see that (123) is implied by

$$\left|\sum_{i\in S_o} q_{ij}\ z^i+b^j-A^j y\right| \leq (<) \frac{s-l_o}{\alpha_j} \frac{Q_x}{\max\limits_{i\in S_o}\left|\frac{z^i-\tilde z}{\tilde\alpha_i\ c^o_{ij}}\right|}, \tag{124}$$

where the right hand side of (124) can also be replaced by its lower bound

$$\frac{s-l_o}{\alpha_j} \frac{Q_x}{\max\limits_{i\in S_o} \frac{||A^i-\tilde A||\cdot||x||+|b^i-\tilde b|}{\tilde\alpha_i\ c^o_{ij}}} .$$

If $m=1$, $\alpha_j = \frac{1}{r}$ for all $j=1,2,...,r$ and (102.1)-(102.3) hold true, then, cf. § 8.2, $c^o_{ij} = \frac{1}{r-l_o}$, $\sum_{i\in S_o} q_{ij}\ z^i = \frac{1}{r-l_o} \sum_{j\in R_o} (A^j y-b^j)$,

and (123) is reduced to

$$(s-l_o)\frac{r}{r-l_o} \cdot \frac{Q_x}{\min\limits_{i\in S_o} \frac{z^i-\tilde z}{\tilde\alpha_i}} \leq (<)((\frac{1}{r-l_o} \sum_{j\in R_o} A^j)-A^j)y-((\frac{1}{r-l_o} \sum_{j\ R_o} b^j)-b^j))$$

$$\leq (<)(s-l_o)\frac{r}{r-l_o} \cdot \frac{Q_x}{\max\limits_{i\in S_o} \frac{z^i-\tilde z}{\tilde\alpha_i}} . \tag{125}$$

Inequality (125) is implied by

$$|((\frac{1}{r-l_o}\sum_{j\in R_o} A^j)-A^j)y-((\frac{1}{r-l_o}\sum_{j\in R_o} b^j)-b^j)| \leq (<) \frac{r(s-l_o)}{r-l_o} \cdot \frac{Q_x}{\max\limits_{i\in S_o} \frac{|z^i-\tilde{z}|}{\tilde{\alpha}_i}} . \tag{126}$$

If m=1, $b^j=b_o$ for all j=1,2,...,r, and (102.1)-(102.3) hold true, then (123) has the form

$$\frac{s-l_o}{\alpha_j} \frac{Q_x}{\min\limits_{i\in S_o} \frac{(A^i-\tilde{A})x}{\tilde{\alpha}_i\ c^o_{ij}}} \leq (<) (1-\hat{\alpha}_j \sum_{i\in S_o} \tilde{\alpha}_i)\tilde{A}x + \hat{\alpha}_j(\sum_{j\in R_o} \alpha_j\ A^j y) - A^j y$$

$$\leq (<) \frac{s-l_o}{\alpha_j} \frac{Q_x}{\max\limits_{i\in S_o} \frac{(A^i-\tilde{A})x}{\tilde{\alpha}_i\ c^o_{ij}}} . \tag{127}$$

Inequality (127) is implied by

$$|(1-\hat{\alpha}_j \sum_{i\in S_o} \tilde{\alpha}_i)\tilde{A}x+\hat{\alpha}_j \sum_{k\in R_o} \alpha_k\ A^k y-A^j y| \leq (<) \frac{s-l_o}{\alpha_j} \frac{Q_x}{||x||} \cdot \frac{1}{\max\limits_{i\in S_o} \frac{||A^i-\tilde{A}||}{\tilde{\alpha}_i\ c^o_{ij}}} . \tag{128}$$

If, in addition, $\alpha_j=\frac{1}{r}$ for all $j \in R$, then (128) is equivalent to

$$|((\frac{1}{r-l_o}\sum_{j\in R_o} A^j)-A^j)y| \leq (<) \frac{r(s-l_o)}{r-l_o} \frac{Q_x}{||x||} \frac{1}{\max\limits_{i\in S_o} \frac{||A^i-\tilde{A}||}{\tilde{\alpha}_i}} . \tag{129.1}$$

Finally, if also $l_o=0$, then (129.1) is reduced to

$$|(\bar{A}-A^j)y| \leq (<)\ s \frac{Q_x}{||x||} \frac{1}{\max\limits_{i\in S_o} \frac{||A^i-\tilde{A}||}{\tilde{\alpha}_i}} . \tag{129.2}$$

For a further discussion of the above inequalities we have to study Q_x in more detail. For m=1 it is, cf. (95.3) and (103),

$$Q_x = \frac{1}{s-l_o} \sum_{i\in S_o} (z^i-\tilde{z})^2 = \frac{1}{s-l_o} \sum_{i\in S_o} ((A^i-\tilde{A})x-(b^i-\tilde{b}))^2 =$$

$$= x'Qx - 2x'(\frac{1}{s-l_o} \sum_{i\in S_o} (A^i-\tilde{A})'(b^i-\tilde{b})) + \frac{1}{s-l_o} \sum_{i\in S_o} (b^i-\tilde{b})^2,$$

where $Q=Q(S_o)$ is the n x n matrix

$$Q = \frac{1}{s-l_o} \sum_{i\in S_o} (A^i-\tilde{A})'(A^i-\tilde{A}). \tag{130}$$

We remember that $Q_x>0$ if and only if $s-l_o>1$. If

$$\rho(x) = \rho(x|S_o) = \frac{x'Qx}{||x||^2} \tag{131.1}$$

is the Rayleigh quotient, and

$$\gamma = \gamma(S_o) = \inf_{||x||=1} x'Qx \tag{131.2}$$

denotes the smallest eigenvalue of Q, and $\mu = \mu(S_o)$, $\delta = \delta(S_o)$ are defined by

$$\mu = ||p||,\ p = \frac{1}{s-l_o} \sum_{i\in S_o} (A^i-\tilde{A})'(b^i-\tilde{b}), \tag{131.3}$$

$$\delta = \frac{1}{s-l_o} \sum_{i\in S_o} (b^i-\tilde{b})^2, \tag{131.4}$$

then for Q_x we have the lower bounds

$$Q_x \geq \rho(x)||x||^2 - 2\mu||x|| + \delta \geq \gamma||x||^2-2\mu||x||+\delta. \tag{132}$$

If $b^i=b_o$ for all $i \in R$, where b_o is a fixed number, then $\mu = \delta = 0$ and (129) reads

$$Q_x \geq \rho(x)||x||^2 \geq \gamma||x||^2. \tag{132.1}$$

According to (130) and (131.2) it is $\gamma>0$ if and only if there is no vector $x\neq 0$ such that

$A^i x = A^j x$ for all $i,j \in S_o$.

Hence, we have the following lemma:

Lemma 8.2. It is $\gamma>0$ if and only if there is an index $i_o \in S_o$ such that

$$\text{rank}(A^i - A^{i_o})_{i \in S_o} = n, \tag{133.1}$$

where $(A^i - A^{i_o})_{i \in S_o}$ denotes the $|S_o| \times n$ matrix having the rows $A^i - A^{i_o}, i \in S_o$.

On the other hand, from (130) and (131.2) we obtain

$$\gamma = \min_{||x||=1} \frac{1}{s-l_o} \sum_{i \in S_o} ((A^i - \tilde{A})x)^2 \leq \frac{1}{s-l_o} \sum_{i \in S_o} ||A^i - \tilde{A}||^2 \leq$$

$$\leq \max_{i \in S_o} ||A^i - \tilde{A}||^2. \tag{133.2}$$

Note.

Condition (133.1) obviously implies that $s-l_o>n$. Hence, (133.1) is always violated if $r\leq n$. However, if $r\leq n$, then by Theorem 7.2.1 we know that for every vector $x \in \mathbb{R}^n$ (12.1)-(12.3) has nontrivial solutions (y,T), i.e. $T \neq T^o$, cf. (14), and $A^j y \neq A^j x$ for at least one $j \in R$ provided that only rank $\mathbb{A}=r$, where $\mathbb{A}$ is the $\mathbf{r} \times n$ matrix having the rows A^i, $i \in R$.

Suppose now that $\gamma>0$, hence $r>n$. According to (132) we then know that the inequality $Q_x>0$ holds for all vectors $x \in \mathbb{R}^n$ such that

$$||x|| \geq 0, \text{ if } \mu^2 - \gamma\delta < 0 \tag{134}$$

$$||x|| > \frac{\mu}{\gamma} + \frac{1}{\gamma}\sqrt{\mu^2 - \gamma\delta}, \text{ if } \mu^2 - \gamma\delta \geq 0.$$

If x satisfies (134), then it is $Q_x>0$ and because of (132) we find that inequality (121) is implied by

$$(z^i-\tilde{z})(\sum_{i\in S_o} q_{ij} z^i+b^j-A^jy)\leq(<)\frac{(s-l_o)\tilde{\alpha}_i}{\alpha_j} c^o_{ij}(\rho(x)||x||^2-2\mu||x||+\delta), \tag{135}$$

where $\sum_{i\in S_o} q_{ij} z^i$ is given by (104).

8.3.2. The numerical stationarity criterion (114). Considering Corollary 8.1, we suppose now that $\gamma>0$ and $c^o_{ij}>0$ for all $i\in S_o$, $j\in R_o$. We first note that according to (132) the regularity condition in the definition (108) of $X(p)$ can be replaced by the simple norm condition (134). Since inequality (112.3a) can be replaced here by (121), (135), resp., with the strict inequality sign, Corollary 8.1 holds also if the functions $J_1(x)$, $J_1^q(x)$, $J_2(x)$ on $X(p)$, cf. (107), (108) and (113), are defined by

$$J_1(x) = \min_{\substack{\bar{A}y=\bar{a}\\ A^{j_l}y-b^{j_l}=a^{i_l}-b^{i_l}\\ l=1,\ldots,l_o}} \max_{j\in R_o} \alpha_j\left|\sum_{i\in S_o} q_{ij} z^i+b^j-A^jy\right|, \tag{136.1}$$

$$J_1^q(x) = \min_{\substack{\bar{A}y=\bar{a}\\ A^{j_l}y-b^{j_l}=a^{i_l}-b^{i_l},\\ l=1,\ldots,l_o}} \sqrt{\sum_{j\in R_o} \alpha_j^2(\sum_{i\in S_o} q_{ij} z^i+b^j-A^jy)^2}, \tag{136.1a}$$

where $J_1(x) \leq J_1^q(x) \leq \sqrt{r-l_o}\, J_1(x)$ for all $x\in X(p)$; furthermore,

$$J_2(x) = \hat{Q}_x \min_{\substack{i\in S_o\\ j\in R_o}} \frac{(s-l_o)\tilde{\alpha}_i\, c^o_{ij}}{|z^i-\tilde{z}|} = \frac{\hat{Q}_x}{\max\limits_{\substack{i\in S_o\\ j\in R_o}} \frac{|z^i-\tilde{z}|}{(s-l_o)\tilde{\alpha}_i\, c^o_{ij}}}. \tag{136.2}$$

Here it is $\hat{Q}_x=Q_x$, or $\hat{Q}_x$ denotes one of the following lower bounds of Q_x

$$Q_x \geq \hat{Q}_x = \rho(x)||x||^2 - 2\mu||x|| + \delta,$$
$$Q_x \geq \hat{Q}_x = \gamma||x||^2 - 2\mu||x|| + \delta, \tag{136.3}$$

resp., where $\rho(x) \geq \gamma, \mu, \delta$ are defined by (131).

Let us now consider the important special case

$l_o = 0$ and $\alpha_j = \frac{1}{r}$ for all $j \in R$

in more detail. According to (107) the parameter family p is then given by

$$p = (S, (J_i)_{i \in S}, \bar{a}), \tag{137.1}$$

and X(p) is defined, see (108),(109), by

$$X(p) = \{x \in H(S, (J_i)_{i \in S}) : \bar{A}x = \bar{a},\ \hat{Q}_x > 0\}. \tag{137.2}$$

Of course, if $\bar{A}=0$, then the parameter $\bar{a}$ in (137.1) and the equality constraint $\bar{A}x=\bar{a}$ in (137.2) are cancelled. In (136.1) and (136.1a) we have the constraint $\bar{A}y=\bar{a}$, hence, using (115) and (108) it is

$$\sum_{i \in S} q_{ij} z^i = \bar{A}x - \bar{b} = \bar{a} - \bar{b} = \bar{A}y - \bar{b}.$$

Since $c^o_{ij} = \frac{1}{r}$ for all $i \in S, j \in R$, we now find

$$J_1(x) = \min_{\bar{A}y=\bar{a}} \frac{1}{r} \max_{j \in R} |(\bar{A} - A^j)y - (\bar{b} - b^j)| \tag{138.1}$$

$$J_1^q(x) = \min_{\bar{A}y=\bar{a}} \sqrt{\frac{1}{r^2} \sum_{j \in R} ((\bar{A} - A^j)y - (\bar{b} - b^j))^2} \tag{138.1a}$$

as well as

$$J_2(x) = \frac{\hat{Q}_x}{\max_{i \in S} \frac{r}{s\,\tilde{\alpha}_i} |z^i - \tilde{z}|}. \tag{138.2}$$

We observe that $J_1(x)$, $J_1^q(x)$ are constants if $\bar{A}=0$. Moreover, if $\bar{A} \neq 0$, then $J_1(x)$, $J_1^q(x)$ depend only on $\bar{a}$. It is easy to see that

$$J_1(x) \leq \tilde{J}_1(x),\ J_1(x) \leq J_1^q(x) \leq \sqrt{r}\, J_1(x) \leq \sqrt{r}\, \tilde{J}_1(x) \text{ and}$$

$$J_2(x) \geq \tilde{J}_2(x) \text{ for all } x \in X(p)$$

if the functions $\tilde{J}_1(x)$, $\tilde{J}_2(x)$ on $X(p)$ are defined by

$$\tilde{J}_1(x) = \min_{\bar{A}y=\bar{a}} \frac{1}{r} (\max_{j \in R} ||\bar{A}-A^j||)||y|| + \frac{1}{r} \max_{j \in R} |\bar{b}-b^j| =$$

$$= \begin{cases} \frac{1}{r} \frac{|\bar{a}|}{||\bar{A}||} \max_{j \in R} ||\bar{A}-A^j|| + \frac{1}{r} \max_{j \in R} |\bar{b}-b^j|, & \text{if } \bar{A} \neq 0 \\ \frac{1}{r} \max_{j \in R} |\bar{b}-b^j|, & \text{if } \bar{A}=0, \end{cases} \tag{139.1}$$

$$\tilde{J}_2(x) = \frac{1}{r} \frac{\hat{Q}_x}{||x|| \max_{i \in S} \frac{||A^i-\bar{A}||}{s\, \tilde{\alpha}_i} + \max_{i \in S} \frac{|b^i-\bar{b}|}{s\, \tilde{\alpha}_i}}, \tag{139.2}$$

where again $\hat{Q}_x$ denotes one of the lower bounds of Q_x defined in (136.3). Consequently, from Corollary 8.1, we obtain this result:

Corollary 8.3. Let $x \in \overset{o}{D} \cap X(p)$ for some parameter family p. a) If $J_1(x) = 0$ or $J_1^q(x) = 0$ or $\tilde{J}_1(x) = 0$, then x is not D-stationary. b) If x is D-stationary, then each of the following inequalities hold true:

$$\tilde{J}_1(x) \geq J_1(x) \geq J_2(x) \geq \tilde{J}_2(x), \tag{140.1}$$

$$\sqrt{r}\, \tilde{J}_1(x) \geq \sqrt{r}\, J_1(x) \geq J_1^q(x) \geq J_1(x) \geq J_2(x) \geq \tilde{J}_2(x), \tag{140.2}$$

especially, it is

$$\left(\frac{|\bar{a}|}{||\bar{A}||} \max_{j \in R} ||\bar{A}-A^j|| + \max_{j \in R} |\bar{b}-b^j|\right)\left(||x|| \max_{i \in S} \frac{||A^i-\tilde{A}||}{s\, \tilde{\alpha}_i} + \max_{i \in S} \frac{|b^i-\tilde{b}|}{s\, \tilde{\alpha}_i}\right)$$

$$\geq Q_x \geq \rho(x)||x||^2-2\mu||x||+\delta \geq \gamma||x||^2-2\mu||x||+\delta, \tag{140.3}$$

where the term $\frac{|\bar{a}|}{||\bar{A}||} \max_{j \in R} ||\bar{A}-A^j||$ is cancelled for $\bar{A}=0$.

Note

Inequality (140.3) implies that for $\gamma>0$ all $x \in \overset{o}{D} \cap X(p)$ having a sufficiently large norm $||x||$ are not D-stationary.

Let us now study the case

$l_o=0$ and $\alpha_j=\frac{1}{r}$, $b^j=b_o$ for all $j \in R$,

where b_o is a fixed number. Here it is $\mu=\delta=0$, and (137.2) yields

$$X(p) = \{x \in H(S,(J_i)_{i \in S}): \bar{A}x=\bar{a}, \rho(x)>0\}, \tag{141}$$

where the constraint $\bar{A}x=\bar{a}$ is cancelled for $\bar{A}=0$.

If $\bar{A}=0$, then $J_1(x)=J_1^q(x)=\tilde{J}_1(x)=0$ for all $x \in X(p)$, see Corollary 8.3. If $\bar{A}\neq 0$, then

$$J_1^q(x) = \frac{1}{\sqrt{r}} \frac{|\bar{a}|}{\sqrt{\bar{A}\,\tilde{Q}^{-1}\bar{A}'}} \text{ with } \tilde{Q} = \frac{1}{r} \sum_{j \in R} (\bar{A}-A^j)'(\bar{A}-A^j),$$

$$\tilde{J}_1(x) = \frac{1}{r} \frac{|\bar{a}|}{||\bar{A}||} \max_{j \in R} ||\bar{A}-A^j||.$$

Therefore, Corollary 8.3 yields this next

Corollary 8.4. a) If $\bar{A}=0$ or $\bar{A}\neq 0$ and $\bar{a}=0$, then $\overset{o}{D} \cap X(p)$ contains no D-stationary points. b) Let $\bar{A}\neq 0$. If $x \in \overset{o}{D} \cap X(p)$ is D-stationary, then $\bar{a}\neq 0$ and

$$\min_{\bar{A}y=\bar{a}} \max_{j \in R} |(\bar{A}-A^j)y| \geq \frac{\rho(x)||x||^2}{\max_{i \in S} \frac{1}{s\,\tilde{\alpha}_i} |(A^i-\tilde{A})x|} \tag{142.1}$$

as well as

$$C \frac{|\bar{a}|}{||\bar{A}||} \geq \rho(x)||x|| \geq \gamma||x||, \tag{142.2}$$

where the constant $C>0$ is defined by

$$C = \max_{\substack{i \in S \\ j \in R}} ||\bar{A}-A^j|| \frac{1}{s\,\tilde{\alpha}_i} ||A^i-\tilde{A}||. \tag{142.3}$$

Note. According to (133.2) we know that

$C \geq \gamma$ for $S = R$.

Suppose now that $\bar{A}\neq 0$. Then $\frac{|\bar{a}|}{||\bar{A}||}$ is the minimal norm of an element x contained in the linear manifold $L(\bar{a}) = \{x \in \mathbb{R}^n: \bar{A}x = \bar{a}\}$. Since X(p) is a subset of $L(\bar{a})$, the inequalities in Corollary 8.4 mean that the D-stationary points contained in $\overset{o}{D} \cap \{x \in H(S,(J_i)_{i\in S}): \rho(x)>0\}$ are also elements of every cone K_1,K_2,K_3 defined by the following inequalities

$$\min_{\bar{A}y=\bar{A}x} \max_{j\in R} |(\bar{A}-A^j)y| \geq \frac{\rho(x)||x||^2}{\max\limits_{i\in S} \frac{1}{s\,\tilde{\alpha}_i} |(A^i-\tilde{A})x|}, \tag{143.1}$$

$$C\,\frac{|\bar{A}x|}{||\bar{A}||} \geq \rho(x)||x||, \tag{143.2}$$

$$C\,\frac{|\bar{A}x|}{||\bar{A}||} \geq \gamma||x||, \text{ respectively.} \tag{143.3}$$

8.3.3. <u>Solving (102.1)-(102.4) for x with given y.</u> Further theorems for the existence of solutions (y,T) of the relations (12.1)-(12.4) may be obtained by exchanging the role of the vectors x and y. Hence, as already mentioned in remark (c) after the proof of Theorem 8.1, for a given $y \in \mathbb{R}^n$ or $y \in D$ the system of relations (102) is now interpreted as a condition for the n-vector x. Note, that (102) can also be interpreted as a condition for the pair (x,y). We recall that for $x \in \mathbb{R}^n$ the index sets S_x, J_{xi}, $i \in S_x$, are defined by (90).

Let y be a given n-vector. Put $l_o=0$ for simplification. Given an index set $S \subset R$ with $1 \in S$ and a partition $J_i, i \in S$, of R such that $i \in J_i$ for every $i \in S$, solutions x of (102) are sought which lie in the set $H(S,(J_i)_{i\in S})$ defined by (109). Hence, x satisfies the relations $S_x=S$ and $J_{xi}=J_i$ for every $i \in S$. Since $l_o=0$, it is $S_o=S$ and $R_o=R$. Define

$$\tilde{\alpha}_i = \sum_{t \in J_i} \alpha_t \text{ for every } i \in S,$$

and assume that $c^o_{ij}>0$ for every $i \in S, j \in R$, where c^o_{ij} is given by (93.2). In order to describe further constraints for x, we need the following linear manifold

$$L(\bar{a},\tilde{a}) = \begin{cases} \{x \in \mathbb{R}^n:\ \bar{A}x=\bar{a},\ \tilde{A}x=\tilde{a}\}, \text{ if } \bar{A}\neq 0 \text{ and } \alpha_j \neq \frac{1}{r} \text{ for all } j \in R \\ \{x \in \mathbb{R}^n:\ \tilde{A}x=\tilde{a}\}, \text{ if } \bar{A}=0 \text{ and } \alpha_j \neq \frac{1}{r} \text{ for all } j \in R \\ \{x \in \mathbb{R}^n:\ \bar{A}x=\bar{a}\}, \text{ if } \bar{A}\neq 0 \text{ and } \alpha_j = \frac{1}{r} \text{ for all } j \in R \\ \mathbb{R}^n, \text{ if } \bar{A}=0 \text{ and } \alpha_j = \frac{1}{r} \text{ for all } j \in R, \end{cases} \quad (144)$$

where $\tilde{A}$ is defined by (103) and $\bar{a},\tilde{a}$ are given fixed numbers. Suppose that $(\bar{a},\tilde{a}) \in \mathbb{R}^2$ is selected such that $L(\bar{a},\tilde{a})$ is an unbounded set in $\mathbb{R}^n$. This is a very week assumption; indeed, $L(\bar{a},\tilde{a})$ is an unbounded linear manifold for every $(\bar{a},\tilde{a}) \in \mathbb{R}^2$ if rank $\begin{pmatrix}\bar{A}\\ \tilde{A}\end{pmatrix}=2<n$, rank $\tilde{A}=1<n$, rank $\bar{A}=1<n$, and for $\bar{A}=0$, $\alpha_j = \frac{1}{r}$ for all $j \in R$, respectively.

Due to the definition (109) of $H(S,(J_i)_{i \in S})$, for every $x \in L(\bar{a},\tilde{a})$ there is a subset $S \subset R$ with $1 \in S$ and a partition $J_i, i \in S$, of R with $i \in J_i$ for every $i \in S$ such that x is an element of

$$V(S,(J_i)_{i \in S},\bar{a},\tilde{a}) = H(S,(J_i)_{i \in S}) \cap L(\bar{a},\tilde{a}). \quad (145)$$

Since $L(\bar{a},\tilde{a})$ is unbounded by assumption and there is only a finite number of intersections of the type (145), we may select $S \subset R$ with $1 \in S$ and the partition $(J_i)_{i \in S}$ of R with $i \in J_i$ for every $i \in S$ such that $V(S,(J_i)_{i \in S},\bar{a},\tilde{a})$ is an unbounded subset of $\mathbb{R}^n$.

For $S=\{1\}$, $J_1=R$ we have that

$$H(\{1\},\{R\}) = \{x \in \mathbb{R}^n:\ (A^i-A^j)x=b^i-b^j \text{ for all } i,j \in R\}.$$

Hence, we may assume - without essential restrictions - that $|S|>1$. If $S\neq R$, then (111.1) yields

$$V(S,(J_i)_{i\in S},\bar{a},\tilde{a}) \subset \bigcap_{j\in J_i, i\in S} (L(\bar{a},\tilde{a})\cap H^{ij});$$

moreover, in the case $S=R$, $J_i=\{i\}, i\in R$, by (111.2) it is

$$V(S,(J_i)_{i\in S},\bar{a},\tilde{a}) = L(\bar{a},\tilde{a})\setminus \bigcup_{i\neq j}(L(\bar{a},\tilde{a})\cap H^{ij}) =$$

$$= \bigcap_{i\neq j}(L(\bar{a},\tilde{a})\setminus H^{ij}),$$

where H^{ij} is defined by (110). Consequently, the choice $S=R$, $J_i=\{i\}, i\in R$, yields an unbounded set $V(R,(\{i\})_{i\in R}, \bar{a},\tilde{a})$ if $L(\bar{a},\tilde{a})\cap H^{ij}, i,j\in R, i\neq j$, are proper linear submanifolds of $L(\bar{a},\tilde{a})$. This holds if for $i,j\in R, i\neq j$, we have that $L(\bar{a},\tilde{a})\cap H^{ij}=\emptyset$, or $L(\bar{a},\tilde{a})\cap H^{ij}$ is a singleton, or there is no 2x2 matrix Λ_{ij}, number Λ_{ij}, resp., such that $A^i-A^j=\Lambda_{ij}\binom{\bar{A}}{\tilde{A}}$, $A^i-A^j=\Lambda_{ij}\,\tilde{A}$, $A^i-A^j=\Lambda_{ij}\,\bar{A}$, resp., or $L(\bar{a},\tilde{a}) = \mathbb{R}^n$.

Note. If $A^i-A^j=\Lambda_{ij}\binom{\bar{A}}{\tilde{A}}$ for every $i\neq j$ with some 2x2 matrix Λ_{ij}, then $A^i=\Lambda_i\binom{\bar{A}}{\tilde{A}}$ with some 2x2 matrix Λ_i for every $i\in R$, and conversely. A corresponding statement holds if $A^i-A^j=\Lambda_{ij}\,\bar{A}$, $A^i-A^j=\Lambda_{ij}\,\tilde{A}$, resp., with some number Λ_{ij} for every $i\neq j$.

Proceeding the construction of a solution x of (102) for a given vector y, we are now going to solve inequality (121), (135), resp., for x on the intersection $V=V(S,(J_i)_{i\in S},\bar{a},\tilde{a})$ defined by (145), where we suppose that $S,(J_i)_{i\in S},\bar{a},\tilde{a}$ are selected such that V is unbounded.

According to (104.4) and (104.6) for every $x\in V$ it is

$$\sum_{i\in S} q_{ij}\, z^i=q_j(\bar{a},\tilde{a}) = \begin{cases} (1-\hat{\alpha}_j)(\tilde{a}-\tilde{b})+\hat{\alpha}_j(\bar{a}-\bar{b}), & \text{if } \bar{A}\neq 0 \text{ and } \alpha_j\neq \frac{1}{r} \text{ for all } j\in R \\ (1-\hat{\alpha}_j)(\tilde{a}-\tilde{b})-\hat{\alpha}_j\bar{b}, & \text{if } \bar{A}=0 \text{ and } \alpha_j\neq \frac{1}{r} \text{ for all } j\in R \\ \bar{a}-\bar{b}, & \text{if } \bar{A}\neq 0 \text{ and } \alpha_j= \frac{1}{r} \text{ for all } j\in R \\ -\bar{b}, & \text{if } \bar{A}=0 \text{ and } \alpha_j= \frac{1}{r} \text{ for all } j\in R, \end{cases} \tag{146}$$

where $\hat{\alpha}_j$ is defined by (104.2). Inserting now (146) into (121), on V inequality (121) reads

$$((A^i-\tilde{A})x-(b^i-\tilde{b}))(q_j(\bar{a},\tilde{a})+b^j-A^jy)\leq(<) \frac{s\,\tilde{\alpha}_i}{\alpha_j} c^o_{ij}(x'Qx-2p'x+\delta), \quad (147)$$

where Q,p,δ are defined as in (130) and (131). We remember, that in accordance with (102.2), the n-vectors x,y and the number $\bar{a}$ must be related by

$$\bar{a} = \bar{A}x = \bar{A}y. \quad (148)$$

Let $p_{ij} \in \mathbb{R}^n$ and $\delta_{ij} \in \mathbb{R}$ be defined by

$$p_{ij} = p + \frac{1}{2}\,\frac{\alpha_j}{s\,\tilde{\alpha}_i\,c^o_{ij}}(q_j(\bar{a},\tilde{a})+b^j-A^jy)(A^i-\tilde{A})', \quad (149.1)$$

$$\delta_{ij} = \delta + \frac{\alpha_j}{s\,\tilde{\alpha}_i\,c^o_{ij}}\,(q_j(\bar{a},\tilde{a})+b^j-A^jy)(b^i-\tilde{b}). \quad (149.2)$$

Then (147) has the form

$$x'Qx - 2p_{ij}'x + \delta_{ij} \geq(>)0. \quad (150)$$

If Q is regular, then (150) is equivalent to

$$(x-Q^{-1}p_{ij})'Q(x-Q^{-1}p_{ij}) \geq(>)\ p_{ij}'Q^{-1}p_{ij}-\delta_{ij}. \quad (150a)$$

Hence, for given vector y, on the unbounded set $V=V(S,(J_i)_{i\in S}, \bar{a},\tilde{a})$ inequality (150), and therefore also (121), turns out to be a quadratic condition for x.

Looking for sufficient conditions for (150), we first note that (150) is implied by

$$\rho(x)||x||^2-2||p_{ij}||\cdot||x|| + \delta_{ij} \geq(>)0, \quad (151)$$

where $\rho(x)$ is the Rayleigh quotient of Q. Moreover, (151) follows from

$$\gamma||x||^2-2||p_{ij}||\cdot||x|| + \delta_{ij} \geq(>)0, \quad (152)$$

where $\gamma=\gamma(S)$ is defined by (131.2). According to Lemma 8.2 it is $\gamma>0$ if and only if S is selected such that (133.1) holds true, i.e.

$\text{rank}(A^i - A^{i_o})_{i\in S} = n$, where i_o is an arbitrary, but fixed element of S, and $(A^i - A^{i_o})_{i\in S}$ is the sxn matrix having the rows $A^i - A^{i_o}, i \in S$. Defining $\tilde{\mu}=\tilde{\mu}(S,\bar{a},\tilde{a},y), \tilde{\delta}=\tilde{\delta}(S,\bar{a},\tilde{a},y)$ with $\bar{a}=\bar{A}y$, cf. (148), by

$$\tilde{\mu} = \max_{\substack{i\in S\\ j\in R}} ||p_{ij}||, \tag{153.1}$$

$$\tilde{\delta} = \min_{\substack{i\in S\\ j\in R}} \delta_{ij}, \tag{153.2}$$

inequality (151),(152), resp., holds true for all $(i,j) \in S\times R$ if

$$\rho(x)||x||^2 - 2\tilde{\mu}||x|| + \tilde{\delta} \geq (>) 0, \tag{154.1}$$

$$\gamma||x||^2 - 2\tilde{\mu}||x|| + \tilde{\delta} \geq (>) 0, \tag{154.2}$$

respectively. Clearly, for $\gamma>0$ inequality (154.2) holds with the strict inequality sign if

$$||x|| > \begin{cases} 0, & \text{if } \tilde{\mu}^2 - \gamma\tilde{\delta} < 0 \\ \frac{\tilde{\mu}}{\gamma} + \frac{1}{\gamma}\sqrt{\tilde{\mu}^2 - \gamma\tilde{\delta}}, & \text{if } \tilde{\mu}^2 - \gamma\tilde{\delta} \geq 0. \end{cases} \tag{155}$$

Summarizing the above construction, we get the following result:

Theorem 8.3. Let y be a given n-vector. Consider a set $S \subset R$, a partition $J_i, i \in S$, of R and, if not $\alpha_j = \frac{1}{r}$ for all $j \in R$, a number $\tilde{a}$ such that

i) $1 \in S$, $|S|>1$, $i \in J_i$ for every $i \in S$

ii) $c^o_{ij}>0$ for all $i \in S$, $j \in R$ (cf. Lemma 8.1)

iii) $V=V(S,(J_i)_{i\in S},\bar{a},\tilde{a})$, where $\bar{a}=\bar{A}y$, is an unbounded set.

a) Let x be an element of V which fulfills

$$\rho(x)||x||^2 - 2\mu||x|| + \delta > 0, \tag{156}$$

cf. (132), and the quadratic conditions (150) for every $i \in S$, $j \in R$. Then, $Q_x>0$ and (y,T) satisfies (12.1)-(12.3), where $T=(\tau_{ij})$ is given by

$$\tau_{ij} = c^o_{ij} - \frac{\alpha_j}{s\,\tilde{\alpha}_i Q_x}((1-\hat{\alpha}_j)(\tilde{a}-\tilde{b})+\hat{\alpha}_j(\bar{A}y-\bar{b})+b^j-A^j y)(z^i-\tilde{z}) \qquad (157)$$

for every $i \in S, j \in R$. If, in addition to the above assumptions, (150) holds with the strict inequality sign for at least two pairs (i_1,j), $(i_2,j) \in S \times R$, $i_1 \neq i_2$, then (y,T) fulfills (12.1)-(12.4a); if (150) holds always with the strict inequality sign, then (y,T) fulfills (12.1)-(12.4b). The same consequences hold if (150) is replaced by the stronger inequality (151),(152) and (154), respectively.

b) Suppose that, in addition to the above assumptions (i)-(iii), it is

iv) $\gamma=\gamma(S)>0$ (cf. Lemma 8.2).

Let x be an element of V which fulfills the norm conditions

$$\gamma||x||^2-2\mu||x||+\delta>0 \qquad (156a)$$

as well as (152) for every $i \in S, j \in R$. Then $Q_x>0$ and (y,T) satisfies (12.1)-(12.3), where $T = (\tau_{ij})$ is again given by (157). If, in addition to this assumptions, inequality (152) holds true with the strict inequality sign for at least two pairs $(i_1,j),(i_2,j) \in S \times R$, $i_1 \neq i_2$, then (y,T) fulfills (12.1)-(12.4a); if (152) holds always with the strict inequality sign, then (y,T) fulfills also (12.1)-(12.4b). The same consequences are true if (152) is replaced by the stronger inequality (154.2), (155), respectively.

Remark. Suppose that for all $i \in S, j \in R$

$||p_{ij}|| \leq \pi(\tilde{a})$, $\delta_{ij} \geq \delta(\tilde{a})$ for every $y \in D$,

see (149), where $\pi(\tilde{a}),\delta(\tilde{a})$ are certain fixed numbers, then (152) is implied for every $y \in D$ by

$$\gamma||x||^2 - 2\pi(\tilde{a})||x|| + \delta(\tilde{a}) \geq (>) 0.$$

8.3.3.1. Special cases. We consider the important case

$\alpha_j = \frac{1}{r}$ for every $j \in R$.

Then, it is $c^o_{ij} = \frac{1}{r}$, and by (146) we find $q_j(\bar{a},\tilde{a}) = \bar{a}-\bar{b} = \bar{A}y-\bar{b}$. Hence, (149) yields

$$p_{ij} = p + \frac{1}{2\, s\tilde{\alpha}_i} (b^j-\bar{b}-(A^j-\bar{A})y)(A^i-\tilde{A})', \tag{158.1}$$

$$\delta_{ij} = \delta + \frac{1}{s\, \tilde{\alpha}_i} (b^j-\bar{b}-(A^j-\bar{A})y)(b^i-\tilde{b}), \tag{158.2}$$

where p,δ are defined in (131). Furthermore, according to (144) it is $L(\bar{a},\tilde{a}) = \{x \in \mathbb{R}^n: \bar{A}x=\bar{a}\}$ for $\bar{A} \neq 0$ and $L(\bar{a},\tilde{a}) = \mathbb{R}^n$ for $\bar{A}=0$. Hence, (145) and (148) yield

$$V(S,(J_i)_{i \in S},\bar{A}y,\tilde{a}) = \{x \in H(S,(J_i)_{i \in S}): \bar{A}x = \bar{A}y\}. \tag{159}$$

In the special case

$\alpha_j = \frac{1}{r}$ and $b^j = b_o$ for all $j \in R$,

then, by (131), we have that $p=0$, $\mu=0$, $\delta=0$. Hence, (158) yields

$$p_{ij} = \frac{1}{2\, s\tilde{\alpha}_i} (\bar{A}-A^j)y(A^i-\tilde{A})', \quad \delta_{ij} = 0. \tag{160}$$

Consequently, (156), (156a) have the form

$\rho(x)||x||>0$, $\gamma||x||>0$, resp.,

and, because of (160), (151), (152), resp., reads

$$\rho(x)||x|| \geq (>) 2||p_{ij}|| = \frac{1}{s\, \tilde{\alpha}_i} |(A^j-\bar{A})y| \cdot ||A^i-\tilde{A}||, \tag{161.1}$$

$$\gamma||x|| \geq (>) 2||p_{ij}|| = \frac{1}{s\, \tilde{\alpha}_i} |(A^j-\bar{A})y| \cdot ||A^i-\tilde{A}||, \tag{161.2}$$

respectively. Inequalities (161) are implied for all $i \in S, j \in R$ by

$$\rho(x)||x|| \geq (>) C||y||, \tag{162.1}$$

$$\gamma||x|| \geq (>) C||y||, \tag{162.2}$$

resp., where C is again defined by (142.3). If S=R, then $s\cdot\tilde{\alpha}_i=1$, $\tilde{A}=\bar{A}$, and, by inequality (133.2), we know that $C\geq\gamma$.

8.4. The case m>1. After the detailed considerations of the special cases $\alpha_j = \frac{1}{r}$ for all $j\in R$ and m=1 in sections 8.2, 8.3, resp., we now continue the study of the general inequality (102.4), hence,

$$(z^i-\tilde{z})'Q_x^{-1}\Big(\sum_{i\in S_o} q_{ij}\; z^i+b^j-A^jy\Big)\leq(<)\frac{(s-l_o)\tilde{\alpha}_i}{\alpha_j}\; c^o_{ij},$$

where we always suppose that x and y are related according to (102.1)-(102.3).

8.4.1. Solving (102.1)-(102.4) for y with given x. In the first part we consider, corresponding to § 8.3.1, a fixed n-vector x such that $|S_o|=s-l_o>1$ and Q_x is regular, cf. (92) and (96).

If $z^i-\tilde{z}=0$, then the inequality (102.4) holds for all $y\in\mathbb{R}^n$ if and only if $c^o_{ij}\geq(>)0$.

Let now $z^i-\tilde{z}\neq0$ and denote by $z^i_k-\tilde{z}_k$ the k-th component of $z^i-\tilde{z}$. Furthermore, let Q^{-1}_{xk} be the k-th row of Q_x^{-1}. Then (102.4) is implied by

$$Q^{-1}_{xk}\Big(\sum_{i\in S_o} q_{ij}\; z^i+b^j-A^jy\Big)\begin{cases}\leq(<)\dfrac{(s-l_o)\tilde{\alpha}_i}{\alpha_j}\; c^o_{ij}\;\dfrac{z^i_k-\tilde{z}_k}{||z^i-\tilde{z}||^2} & \text{if } z^i_k-\tilde{z}_k>0\\[2ex] \geq(>)\dfrac{(s-l_o)\tilde{\alpha}_i}{\alpha_j}\; c^o_{ij}\;\dfrac{z^i_k-\tilde{z}_k}{||z^i-\tilde{z}||^2}, & \text{if } z^i_k-\tilde{z}_k<0,\end{cases}$$

$$k=1,2,\ldots,m, \tag{163}$$

where for $z^i_k-\tilde{z}_k=0$ there is no constraint, cf. (121a).

We observe that (163) is of the same type as (121a), hence, relations corresponding to (122)-(129) may be derived. Thus, if

$c_{ij}^o \geq 0$ for all $i \in S_o, j \in R_o$, then, see (124), (102.4) is fulfilled for every $i \in S_o, j \in R_o$ if the inequality

$$|Q_{xk}^{-1}(\sum_{i \in S_o} q_{ij} z^i + b^j - A^j y)| \leq (<) \frac{s-1_o}{\alpha_j} \frac{1}{\max\limits_{i \in S_o} \frac{1}{\tilde{\alpha}_i c_{ij}^o} \cdot \frac{||z^i - \tilde{z}||^2}{|z_k^i - \tilde{z}_k|}} \tag{164}$$

holds for every $1 \leq k \leq m$ and $j \in R_o$. There are several relations implying (164). E.g., stronger relations than (164) are obtained if the right hand side of (164) is replaced by its lower bound

$$\frac{s-1_o}{\alpha_j} \cdot \frac{\min\limits_{i \in S_o} |z_k^i - \tilde{z}_k|}{\max\limits_{i \in S_o} \frac{||z^i - \tilde{z}||^2}{\tilde{\alpha}_i c_{ij}^o}}$$

and/or the left hand side of (164) is replaced by one of its upper bounds

$$||Q_{xk}^{-1}|| \cdot ||\sum_{i \in S_o} q_{ij} z^i + b^j - A^j y|| \leq ||Q_x^{-1}|| \cdot ||\sum_{i \in S_o} q_{ij} z^i + b^j - A^j y||.$$

As was mentioned earlier, see (105), in the case $c_{ij}^o \geq 0$ inequality (102.4) is also implied by the norm condition

$$||z^i - \tilde{z}|| \cdot ||Q_x^{-1}|| \cdot ||\sum_{i \in S_o} q_{ij} z^i + b^j - A^j y|| \leq (<) \frac{(s-1_o)\tilde{\alpha}_i}{\alpha_j} c_{ij}^o.$$

Obviously, this inequality, and therefore, also (102.4) hold true for every $(i,j) \in S_o \times R_o$ if

$$||\sum_{i \in S_o} q_{ij} z^i + b^j - A^j y|| \leq (<) \frac{s-1_o}{\alpha_j} \frac{||Q_x^{-1}||^{-1}}{\max\limits_{i \in S_o} \frac{||z^i - \tilde{z}||}{\tilde{\alpha}_i c_{ij}^o}} \tag{165}$$

is true for every $j \in R_o$. For $||Q_x^{-1}|| = \sup\limits_{||w||=1} ||Q_x^{-1} w||$ we have the upper bound

$$||Q_x^{-1}|| \leq \frac{1}{q_x}, \tag{166.1}$$

where $q_x > 0$ is given $(||w|| = ||Q_x Q_x^{-1} w||)$ by

$$q_x = \inf_{||w||=1} ||Q_x w|| = \inf_{||w||=1} w'Q_x w, \tag{166.2}$$

hence, q_x is the minimal eigenvalue of Q_x. According to (95.3) and (103), for $w'Q_x w$ we find

$$w'Q_x w = \frac{1}{s-1_o} \sum_{i \in S_o} (w'(z^i - \tilde{z}))^2 = \frac{1}{s-1_o} \sum_{i \in S_o} (w'((A^i - \tilde{A})x - (b^i - \tilde{b})))^2 =$$

$$= \frac{1}{s-1_o} \sum_{i \in S_o} (w'(A^i - \tilde{A})x)^2 - \frac{2}{s-1_o} \sum_{i \in S_o} w'(A^i - \tilde{A})xw'(b^i - \tilde{b}) +$$

$$+ \frac{1}{s-1_o} \sum_{i \in S_o} (w'(b^i - \tilde{b}))^2 =$$

$$= ||x||^2 \frac{1}{s-1_o} \sum_{i \in S_o} (w'(A^i - \tilde{A})\frac{x}{||x||})^2 - 2||x||\frac{1}{s-1_o} \sum_{i \in S_o} w'(A^i - \tilde{A})\frac{x}{||x||} \cdot$$

$$\cdot w'(b^i - \tilde{b}) + \frac{1}{s-1_o} \sum_{i \in S_o} (w'(b^i - b))^2.$$

Hence, corresponding to (132), for q_x we have

$$q_x \geq \underline{q}_x, \tag{167}$$

where the lower bound $\underline{q}_x$ is defined by

$$\underline{q}_x = \rho(x)||x||^2 - 2\mu(x)||x|| + \delta, \tag{167.1a}$$

$$\underline{q}_x = \rho(x)||x||^2 - 2\bar{\mu}||x|| + \delta, \tag{167.1b}$$

$$\underline{q}_x = \gamma||x||^2 - 2\bar{\mu}||x|| + \delta, \tag{167.1c}$$

respectively. Here, $\rho(x), \gamma, \mu(x), \bar{\mu}$ and δ are defined by

$$\rho(x) = \inf_{||w||=1} \frac{1}{s-1_o} \sum_{i \in S_o} (w'(A^i - \tilde{A})\frac{x}{||x||})^2, \tag{167.2}$$

$$\gamma = \inf_{||x||=1} \rho(x), \tag{167.3}$$

$$\mu(x) = \sup_{||w||=1} \frac{1}{s-1_o} \sum_{i \in S_o} w'(A^i - \tilde{A})\frac{x}{||x||} w'(b^i - \tilde{b}) \tag{167.4}$$

$$\bar{\mu} = \sup_{||x||=1} \mu(x), \tag{167.5}$$

$$\delta = \inf_{||w||=1} \frac{1}{s-1_o} \sum_{i \in S_o} (w'(b^i - \tilde{b}))^2. \tag{167.6}$$

We see that $\rho(x)$ is the minimal eigenvalue of the matrix

$$\frac{1}{s-1_o} \sum_{i \in S_o} (A^i - \tilde{A}) \frac{x}{||x||} \frac{x'}{||x||} (A^i - \tilde{A})'.$$

Furthermore, if $b^j = b_o$ for every $j \in R$, where b_o is a fixed m-vector, then, it is $\mu(x)=0$ for each $x \in \mathbb{R}^n$, $\bar{\mu}=0$ and $\delta=0$.

Corresponding to $\hat{Q}_x$, see (136.3), let $\hat{q}_x$ denote q_x, see (166.2), or one of its lower bounds $\underline{q}_x$, see (167.1a-c), hence $q_x \geq \hat{q}_x$. Under the condition

$$\hat{q}_x > 0 \tag{168}$$

inequality (105), and therefore also (102.4) are implied by

$$||z^i - \tilde{z}|| \cdot || \sum_{i \in S_o} q_{ij} z^i + b^j - A^j y|| \leq (<) \frac{(s-1_o)\tilde{\alpha}_i}{\alpha_j} c^o_{ij} \hat{q}_x, \tag{169}$$

see (166.1). Note, that (169) is related to condition (135) found in the case m=1.

Furthermore, if condition (168) holds, then inequality (165) follows from

$$|| \sum_{i \in S_o} q_{ij} z^i + b^j - A^j y|| \leq (<) \frac{s-1_o}{\alpha_j} \cdot \frac{\hat{q}_x}{\max_{i \in S_o} \frac{||z^i - \tilde{z}||}{\hat{\alpha}_i c^o_{ij}}}. \tag{170}$$

We remember that under the conditions (102.1)-(102.3) it is

$$\sum_{i \in S_o} q_{ij} z^i = (1 - \hat{\alpha}_j \sum_{i \in S_o} \tilde{\alpha}_i)(\tilde{A}x - \tilde{b}) + \hat{\alpha}_j \sum_{j \in R_o} \alpha_j (A^j y - b^j), \tag{171}$$

see (104); in the case $\alpha_j = \frac{1}{r}$ for all $j \in R$, (171) is reduced to

$$\sum_{i\in S_o} q_{ij}\, z^i = \frac{1}{r-l_o} \sum_{j\in R_o} (A^j y-b^j). \tag{171.1}$$

Hence, in the special case

$l_o = 0$ and $\alpha_j = \frac{1}{r}$ for all $j \in R$,

cf. § 8.2, condition (169) has the form

$$||z^i-\tilde{z}||\cdot||(\bar{A}-A^j)y-(\bar{b}-b^j)|| \leq (<)\ s\cdot\tilde{\alpha}_i\cdot\hat{q}_x. \tag{172}$$

8.4.2. The numerical stationarity criterion (114.2). If l_o=o and $\alpha_j=\frac{1}{r}$ for all $j \in R$, then the functions $J_1(x), J_2(x)$ in (114.2):

$J_1(x) \geq J_2(x)$ for a stationary point $x \in D \cap X(p)$

can be defined by

$$J_1(x) = \min_{\bar{A}y=\bar{a}} \max_{j\in R} ||(\bar{A}-A^j)y-(\bar{b}-b^j)|| \tag{173.1}$$

$$J_2(x) = \frac{\hat{q}_x}{\max\limits_{i\in S} \frac{||z^i-\tilde{z}||}{s\,\tilde{\alpha}_i}}. \tag{173.2}$$

Estimates for $J_1(x)$ and $J_2(x)$ from above and below, resp., can be obtained as in (139) for m=1.

In the case

$l_o = 0$ and $\alpha_j = \frac{1}{r}$, $b^j = b_o$ for every $j \in R$,

where b_o is a fixed m-vector, condition (172) reads, see (167),

$$||(A^i-\tilde{A})x||\cdot||(\bar{A}-A^j)y|| \leq (<) s\tilde{\alpha}_i\, \rho(x)||x||^2. \tag{174}$$

Inequality (174) follows from

$$||A^i-\tilde{A}||\cdot||\bar{A}-A^j||\cdot||y|| \leq (<) s\tilde{\alpha}_i\, \rho(x)||x||, \tag{175}$$

cf. (118), (119) and (129) for the case m=1.

8.4.3. Solving (102.1)-(102.4) for x with given y. Corresponding to § 8.3.3 for m=1, we now exchange the role of the vectors x,y in (102), i.e., as mentioned after the proof of Theorem 8.1, for a given vector $y \in \mathbb{R}^n$, $y \in D$, resp., we now interpret (102.1) - (102.4) as a condition for the n-vector x.

As in § 8.3.3, for simplification, we put $l_o=0$, and we assume that $c^o_{ij}>0$ for all $i\in S, j\in R$. For every $x\in\mathbb{R}^n$ the index sets S_x, J_{xi}, $i\in S_x$, are again defined by (90.1), (90.2).

In order to construct for a given vector y solutions x of (102), we consider a system of parameters

$$\tilde{p} = (S,(J_i)_{i\in S},\bar{a},\tilde{a}) \tag{176}$$

composed of an index set $S\subset R$ with $1\in S$, a partition $(J_i)_{i\in S}$ of R with $i\in J_i$ for every $i\in S$, the vector $\bar{a} = \bar{A}y$, and a second m-vector $\tilde{a}$. We then define $H(S,(J_i)_{i\in S})$ again by (109); moreover, the linear manifold $L(\bar{a},\tilde{a})$ is defined by (144), where, unlike § 8.3.3, here $\bar{a},\tilde{a}$ are not numbers, but m-vectors. Finally, given $\tilde{p}$, according to (145), we define

$$V(\tilde{p}) = H(S,(J_i)_{i\in S})\cap L(\bar{a},\tilde{a}).$$

Suppose now that $\tilde{p}$ is selected such that $V(\tilde{p})$ is an unbounded subset of $\mathbb{R}^n$, see the corresponding discussion in § 8.3.3. According to (171), for every $x\in V(\tilde{p})$ it is

$$\sum_{i\in S} q_{ij}\, z^i = q_j(\bar{a},\tilde{a}), \tag{177}$$

where $q_j(\bar{a},\tilde{a})$ is defined as in (146). Inserting now (177) into (169), on $V(\tilde{p})$ inequality (169) reads

$$||(A^i-\tilde{A})x-(b^i-\tilde{b})||\cdot||q_j(\bar{a},\tilde{a})+b^j-A^jy||\le(<)\ \frac{s\ \tilde{\alpha}_i}{\alpha_j}\ c^o_{ij}\ \hat{q}_x \tag{178}$$

provided that $\hat{q}_x>0$, where $\hat{q}_x$ denotes q_x or $\underline{q}_x$, see (166.2), (167.1), respectively. Inequality (178) follows from

$$(||A^i-\tilde{A}||\cdot||x||+||b^i-\tilde{b}||)\cdot||q_j(\bar{a},\tilde{a})+b^j-A^jy||\le(<)\frac{s\ \tilde{\alpha}_i}{\alpha_j}\ c^o_{ij}\ \hat{q}_x. \tag{179}$$

Let $\hat{q}_x=\underline{q}_x$. Using (167.1)-(167.6), we define, cf. (149), μ_{ij} and δ_{ij} by

$$\mu_{ij} = \hat{\mu} + \frac{1}{2}\,\frac{\alpha_j}{s\,\tilde{\alpha}_i\,c^o_{ij}}\,||A^i-\tilde{A}||\cdot||q_j(\bar{a},\tilde{a})+b^j-A^jy||, \tag{180.1}$$

$$\delta_{ij} = \delta - \frac{\alpha_j}{s\,\tilde{\alpha}_i\,c^o_{ij}}\,||b^i-\tilde{b}||\cdot||q_j(\bar{a},\tilde{a})+b^j-A^jy||, \tag{180.2}$$

where $\hat{\mu}$ is defined by (167.4), (167.5), resp., and δ is given by (167.6). Then (179) has the form

$$\rho(x)||x||^2 - 2\mu_{ij}||x|| + \delta_{ij} \geq (>)0 \tag{181.1}$$

or

$$\gamma||x||^2 - 2\mu_{ij}||x|| + \delta_{ij} \geq (>)0, \tag{181.2}$$

cf. (151), (152). Obviously, if $\gamma>0$ and $\hat{\mu}=\bar{\mu}$, then (181.2) is a simple quadratic condition for $||x||$. Defining $\tilde{\mu}$, $\tilde{\delta}$ as in (153) by

$$\tilde{\mu} = \max_{\substack{i\in S\\ j\in R}} \mu_{ij},\quad \tilde{\delta} = \min_{\substack{i\in S\\ j\in R}} \delta_{ij}, \tag{182}$$

(181.1), (181.2), resp., holds for all $i\in S, j\in R$ if

$$\rho(x)||x||^2 - 2\tilde{\mu}||x|| + \tilde{\delta} \geq (>)0, \tag{183.1}$$

$$\gamma||x||^2 - 2\tilde{\mu}||x|| + \tilde{\delta} \geq (>)0, \tag{183.2}$$

resp., cf. (154) for m=1. Again, if $\gamma>0$ and $\hat{\mu}=\bar{\mu}$, then (183.2) is a quadratic condition for $||x||$.

Corresponding to Theorem 8.3 for m=1, in the case m>1 we now obtain the following result:

Theorem 8.4. Let y be a given n-vector. Consider a set $S\subset R$, a partition $J_i, i\in S$, of R and, if not $\alpha_j = \frac{1}{r}$ for all $j\in R$, an m-vector $\tilde{a}$ such that

i) $1\in S, |S|>1, i\in J_i$ for every $i\in S$

ii) $c^o_{ij}>0$ for all $i\in S, j\in R$

iii) $V = V(S,(J_i)_{i\in S},\bar{a},\tilde{a})$, $\bar{a}=\bar{A}y$, is an unbounded set.

a) Let x be an element of V such that condition (168) holds true and (181.1), (181.2), resp., is fulfilled for every $(i,j) \in S \times R$. Then Q_x is regular, and (y,T) satisfies (12.1)-(12.3), where $T=(\tau_{ij})$ is given by

$$\tau_{ij} = c^o_{ij} - \frac{\alpha_j}{s\,\tilde{\alpha}_i}(z^i-\tilde{z})'Q_x^{-1}((1-\hat{\alpha}_j)(\tilde{a}-\tilde{b})+\hat{\alpha}_j(\bar{A}y-\bar{b})+b^j-A^jy) \qquad (184)$$

for every $i \in S, j \in R$. If, in addition to the above assumptions, (181.1), (181.2), resp., holds with the strict inequality sign for at least two pairs $(i_1,j),(i_2,j) \in S \times R$, $i_1 \neq i_2$, then (y,T) fulfills (12.1)-(12.4a); if (181.1), (181.2), resp., holds always with the strict inequality sign, then (y,T) fulfills (12.1)-(12.4b).

b) Suppose that in addition to the above assumptions (i)-(iii) it is

iv) $\gamma > 0$.

Let x be an element of V which fulfills the quadratic norm conditions

$$\gamma||x||^2 - 2\bar{\mu}||x||+\delta>0 \qquad (185)$$

and (181.2) with $\hat{\mu}=\bar{\mu}$ for every $i \in S, j \in R$. Then Q_x is regular and (y,T) satisfies (12.1)-(12.3), where $T=(\tau_{ij})$ is given again by (184). If, in addition to the above assumptions, (181.2) holds with the strict inequality sign for at least two pairs $(i_1,j),(i_2,j) \in S \times R$, $i_1 \neq i_2$, then (y,T) fulfills (12.1)-(12.4a); if (181.2) holds always with the strict inequality sign, then (y,T) fulfills (12.1)-(12.4b). The same consequences hold if (181.2) is replaced by the stronger condition (183.2).

According to Theorem 8.4, the main problem is to find n-vectors x satisfying (168), i.e. $\hat{q}_x>0$. Since $\hat{q}_x$ denotes q_x or $\underline{q}_x$, according to (167), (167.1) it is

$$\hat{\underline{q}}_x \geq \rho(x)||x||^2 - 2\bar{\mu}||x||+\delta \geq \gamma||x||^2 - 2\bar{\mu}||x||+\delta.$$

Consequently, we have to find sufficient conditions for $\rho(x)>0$, $\gamma>0$, respectively.

a) By the definition (167.2) of $\rho(x)$ it is $\rho(x)\geq 0$ for all n-vectors $x\neq 0$ and $\rho(x)=0$ holds for an $x\neq 0$ if and only if there is an m-vector $w\neq 0$ such that

$$w'(A^i-\tilde{A})x=0 \text{ for every } i\in S$$

which means that all vectors $A^i x, i\in S$, lie in a certain fixed hyperplane $h^{(m)}$ of $\mathbb{R}^m$. Thus, we have this lemma:

Lemma 8.3. It is $\rho(x)>0$ if and only if not all vectors $A^i x, i\in S$, lie in a certain fixed hyperplane $h^{(m)}$ of $\mathbb{R}^m$. Furthermore, if $S = \{i_1, i_2, \ldots, i_s\}$, then $\rho(x)>0$ if and only if

$$\operatorname{rank}((A^{i_1}-\tilde{A})x, (A^{i_2}-\tilde{A})x, \ldots, (A^{i_s}-\tilde{A})x) \geq m. \tag{186}$$

b) From the definition (167.3) of γ we know that $\gamma\geq 0$, and $\gamma=0$ holds if and only if there are m-, n-vectors $z\neq 0, h\neq 0$ such that

$$z'(A^i-\tilde{A})h=0 \text{ for all } i\in S$$

which means that all matrices $A^i, i\in S$, lie in a fixed hyperplane $h^{(m\cdot n)}$ of $\mathbb{R}^{m\cdot n}$ of the type

$$h^{(m\cdot n)} = \{A\in\mathbb{R}^{m\cdot n}: \text{trace } AB' = \beta\},$$

where $B = zh'$ and $\beta\in\mathbb{R}$. Moreover, $\gamma=0$ is also equivalent to the condition that the positive semidefinite mxm matrix

$$\Gamma(h) = \frac{1}{s}\sum_{i\in S}(A^i-\tilde{A})hh'(A^i-\tilde{A})' = \frac{1}{s}\sum_{i\in S}(A^i-\tilde{A})h((A^i-\tilde{A})h)'$$

is singular for some vector $h\neq 0$. Since $\Gamma(h)$ may also be represented by

$$\Gamma(h) = \begin{pmatrix} h' & 0' & \cdots & 0' \\ 0' & h' & \cdots & 0' \\ \vdots & \vdots & \ddots & \vdots \\ 0' & 0' & \cdots & h' \end{pmatrix} \Gamma_0 \begin{pmatrix} h & 0 & \cdots & 0 \\ 0 & h & \cdots & 0 \\ \vdots & \vdots & \ddots & \vdots \\ 0 & 0 & \cdots & h \end{pmatrix},$$

where Γ_o is the covariance matrix

$$\Gamma_o = \begin{pmatrix} \tilde{cov}(A_1,A_1) & \tilde{cov}(A_1,A_2) & \dots & \tilde{cov}(A_1,A_m) \\ \tilde{cov}(A_2,A_1) & \tilde{cov}(A_2,A_2) & \dots & \tilde{cov}(A_2,A_m) \\ \vdots & \vdots & & \vdots \\ \tilde{cov}(A_m,A_1) & \tilde{cov}(A_m,A_1) & \dots & \tilde{cov}(A_m,A_m) \end{pmatrix}$$

defined by

$$\tilde{cov}(A_k,A_\kappa) = \frac{1}{s} \sum_{i\in S} (A_k^i - A_k)'(A_\kappa^i - A_\kappa), \ k,\kappa=1,\dots,m,$$

we now have the following result:

<u>Lemma 8.4.</u> It is $\gamma>0$ if and only if there is now hyperplane $h^{(m\cdot n)}$ in $\mathbb{R}^{m\cdot n}$ of the type $h^{(m\cdot n)} = \{A \in \mathbb{R}^{m\cdot n}:$ trace $AB' = \beta\}$, where $\beta \in \mathbb{R}$ and $B = zh'$ with m-, n-vectors $z \neq 0$, $h \neq 0$, such that all matrices $A^i, i \in S$, lie in $h^{(m\cdot n)}$. If Γ_o is regular, then $\gamma>0$, where for $m=1$ this condition is also necessary.

A simple consequence of the above lemma is this corollary:

<u>Corollary 8.5.</u> If there is an index $i_o \in S$ such that the set of matrices $\{A^i - A^{i_o} : i \in S\}$ contains $m\cdot n$ linear independent elements, then $\gamma>0$.

Proof. If $\gamma=0$, then, by Lemma 8.4, there is a matrix $B=zh' \neq 0$ such that trace $(A^i - A^{i_o})B' = 0$ for all $i \in S$. Because of the above assumptions this yields the contradiction $B=0$. Hence, we must have $\gamma>0$.

Note. The assumptions in Corollary 8.5 imply that $r \geq m\cdot n$, cf. Theorem 7.2.1.

9. Construction of solutions (y,B) of (46) by using representation (60) of $(A(\omega),b(\omega))$

9.1. System (192.1)-(192.6) for the construction of (y,B). Suppose that the random matrix $(A(\omega),b(\omega))$ is given by (60), i.e. let

$$(A(\omega),b(\omega)) = (A^{(0)},b^{(0)}) + \sum_{t=1}^{L} \xi_t(\omega)(A^{(t)},b^{(t)}),$$

where $(A^{(t)},b^{(t)})$, $t=0,1,\dots,L$ are given fixed $m\times(n+1)$ matrices, and $\xi_1(\omega),\dots,\xi_L(\omega)$ are zero mean random variables having nondegenerated discrete distributions.

In practice, we often find that

$$L < m(n+1) \text{ or even } L \ll m(n+1), \tag{187}$$

thus, the number of generating random variables $\xi_t(\omega)$ in $(A(\omega),b(\omega))$ is much smaller than the total number of matrix elements of $(A(\omega),b(\omega))$.

According to the above assumptions it is $E\xi_t(\omega)=0$ for all $t=1,\dots,L$, hence

$$(\bar{A},\bar{b}) = E(A(\omega),b(\omega)) = (A^{(0)},b^{(0)}).$$

Furthermore, let Ξ denote the $L\times r$ matrix

$$\Xi = \begin{pmatrix} \xi_1^1 & \dots & \xi_1^i & \dots & \xi_1^r \\ \vdots & & \vdots & & \vdots \\ \xi_t^1 & \dots & \xi_t^i & \dots & \xi_t^r \\ \vdots & & \vdots & & \vdots \\ \xi_L^1 & \dots & \xi_L^i & \dots & \xi_L^r \end{pmatrix}$$

of realizations of the discretely distributed random variables $\xi_1(\omega),\dots,\xi_L(\omega)$. Since these random variables are nondegenerated and have zero means, it is

$$V(\xi_t(\omega)) = \sum_{i=1}^{r} (\xi_t^i)^2 \alpha_i > 0, \ t=1,\dots,L, \tag{188.1}$$

$$\Xi_t \alpha = \sum_{i=1}^{r} \xi_t^i \alpha_i = E\,\xi_t(\omega) = 0, \ t=1,\dots,L. \tag{188.2}$$

Here Ξ_t denotes the t-th row of Ξ, and the vector of probabilities $\alpha=(\alpha_1,\dots,\alpha_r)'$ is defined by $\alpha_i=P(\xi_1(\omega)=\xi_1^i,\dots,\xi_L(\omega)=\xi_L^i)$, where $\alpha_i>0$, $i=1,\dots,r$. Supposing that the random variables $\xi_1(\omega),\dots,\xi_L(\omega)$ are defined on the discrete probability space $(\{\omega_1,\dots,\omega_r\}, \mathfrak{A}, P)$, where $\mathfrak{A}$ is the σ-algebra of all subsets of $\{\omega_1,\dots,\omega_r\}$, we may also set $\alpha_i=P(\{\omega_i\})$, $i=1,\dots,r$.

Defining the rxr matrix $B = (b_{ij})$ by $b_{ij} = \frac{\alpha_i \pi_{ij}}{\alpha_j}$, $i,j \in R$, according to § 7.1 we know that (3.1)-(3.3) may be described by (46), thus

$$1_r'B = 1_r', \ B \geq 0$$
$$B\alpha = \alpha$$
$$Z_y = Z_x B,$$

where $1_r = (1,1,\dots,1)' \in \mathbb{R}^r$ and Z_x is the mxr matrix

$$Z_x = (A^1x-b^1, A^2x-b^2,\dots,A^rx-b^r).$$

If rank $\Xi = L$, then for any vectors $x,y \in \mathbb{R}^n$ it is

$$A^j y = A^j x \text{ for all } j \in R \text{ if and only if } A^{(t)}y = A^{(t)}x \text{ for all } t=0,1,\dots,L, \tag{189}$$

cf. § 7.1.6.1.

Having the representation (60) of $(A(\omega),b(\omega))$, equation (46.3) is equivalent to the equations (61.1) and (61.2), i.e.

$$A^{(0)}y = A^{(0)}x$$

$$(A^{(1)}y-b^{(1)},\dots,A^{(L)}y-b^{(L)})\Xi = (A^{(1)}x-b^{(1)},\dots,A^{(L)}x-b^{(L)})\Xi B.$$

Introducing now an auxiliary LxL matrix H, the last equality (61.2) may be split into equations (63.3) and (63.5) which then separates the unknowns y, B from each other. Hence, given $x \in D$, solutions (y,B) of (46) may be obtained by solving for (y,B,H) the following system of relations

$$1_r'B = 1_r', B \geq 0 \qquad (190.1)$$

$$B\alpha = \alpha \qquad (190.2)$$

$$\Xi B = H\Xi \qquad (190.3)$$

$$A^{(0)}y = A^{(0)}x \qquad (190.4)$$

$$(A^{(1)}y-b^{(1)},\dots,A^{(L)}y-b^{(L)}) = (A^{(1)}x-b^{(1)},\dots,A^{(L)}x-b^{(L)})H \qquad (190.5)$$

$$y \in D, \qquad (190.6)$$

cf. (63).The above system (190) can be further simplified if we assume that H is diagonal, thus

$$H = (h_{tt}\ \delta_{t\tau}), \qquad (191)$$

where

$$h = (h_{11}, h_{22}, \dots, h_{LL})'$$

which is the L-vector of diagonal elements of H. Indeed, if (191) holds true, then (190) reads

$$1_r'B = 1_r',\ B \geq 0 \qquad (192.1)$$

$$B\alpha = \alpha \qquad (192.2)$$

$$\Xi_t B = h_{tt}\ \Xi_t,\ t=1,\dots,L \qquad (192.3)$$

$$A^{(0)}y = A^{(0)}x \qquad (192.4)$$

$$A^{(t)}y-b^{(t)}=h_{tt}(A^{(t)}x-b^{(t)}),\ t=1,\dots,L \qquad (192.5)$$

$$y \in D. \qquad (192.6)$$

Remark

a) (192.3) means that Ξ_t is a left-eigenvector of B related to the left-eigenvalue h_{tt} of B. b) Because of (188.2) we find that the tuple

$$B = \alpha 1_r',\ h=0 \qquad (193)$$

satisfies the first three conditions of (192). c) If $A^{(t)}x=b^{(t)}$ for some $1\leq t\leq L$, then (192.5) yields also that $A^{(t)}y=b^{(t)}$.

Let us still mention that, instead of (46), we can also use system (47) taking into consideration that some of the vectors $z^i=A^ix-b^i, i\in R$, may be equal. Hence, corresponding to (46) and (190) we find that solutions $(y,\tilde{B})$ of (47) with $y\in D$ can be obtained by solving for $(y,\tilde{B},H)$ the following system of relations

$$1_s'\tilde{B} = 1_r', \ \tilde{B}\geq 0 \tag{194.1}$$

$$\tilde{B}\alpha = \tilde{\alpha} \tag{194.2}$$

$$\tilde{\Xi}\tilde{B} = H\Xi \tag{194.3}$$

$$A^{(0)}y = A^{(0)}x \tag{194.4}$$

$$(A^{(1)}y-b^{(1)},\ldots,A^{(L)}y-b^{(L)}) = (A^{(1)}x-b^{(1)},\ldots,A^{(1)}x-b^{(1)})H \tag{194.5}$$

$$y\in D. \tag{194.6}$$

We remember that $\tilde{B}$ is an sxr matrix and $\tilde{\alpha}$ is the s-vector having the components $\tilde{\alpha}_i = \sum_{z^l=z^i} \alpha_l$ for $i\in S = \{i_1,i_2,\ldots,i_s\}$. Furthermore $\tilde{\Xi}$ is the Lxs submatrix of Ξ defined by

$$\tilde{\Xi} = \begin{pmatrix} \xi_1^{i_1} & \cdots & \xi_1^{i_\sigma} & \cdots & \xi_1^{i_s} \\ \vdots & & \vdots & & \vdots \\ \xi_t^{i_1} & \cdots & \xi_t^{i_\sigma} & \cdots & \xi_t^{i_s} \\ \vdots & & \vdots & & \vdots \\ \xi_L^{i_1} & \cdots & \xi_L^{i_\sigma} & \cdots & \xi_L^{i_s} \end{pmatrix}.$$

Note. If (191) holds, then a simplification of (194) similar to (192) is obtained.

If $x\in D$ is D-stationary, cf. Definition 4.1 and Theorem 7.4, then system (46), and therefore also the weaker system (190),

(192) have only solutions (y,B), (y,B,H), (y,B,h), respectively, such that $Z_y = Z_x$, or the direction $h=y-x$ is not feasible for D at x. Hence, before we start the construction of solutions (y,B,h) of (192), we first summarize the results on stationary points valid for the present case (60). By Lemma 7.6 and Lemma 7.7 we obtain the following lemma:

Lemma 9.1. a) If $x \in D$ satisfies the linear equations $A^{(t)}x=b^{(t)}$, $t=1,\dots,L$, then x is D-stationary. b) Let rank $\Xi = L$ and suppose that for every $x \in D$ the system of linear equations

$$A^{(0)}y = A^{(0)}x \tag{195}$$
$$A^{(t)}y = b^{(t)}, \quad t=1,\dots,L$$

has a solution $y \in D$, then the set S_D of D-stationary points is given by

$$S_D = \{x \in D: A^{(t)}x = b^{(t)}, \ t=1,\dots,L\}. \tag{196}$$

For $A^{(0)}=0$ we find this corollary:

Corollary 9.1. Let $A^{(0)} = EA(\omega) = 0$. If there is a $y \in D$ such that $A^{(t)}y=b^{(t)}$ for all $t=1,\dots,L$, then S_D is given by (196).

Define now the set H of auxiliary vectors h by

$$H = \{h \in \mathbb{R}^L: \text{there is an } r\text{x}r \text{ matrix B such that } (B,h) \text{ solves } (192.1)\text{-}(192.3)\}. \tag{197}$$

Obviously, H contains $h=0 (B=\alpha 1_r')$ and $h=1_L (B=I)$.
Using (192), solutions (y,B) of (46) may be constructed by the following two-step procedure

- solve (192.1)-(192.3) for (B,h),
- given $h \in H$, solve (192.4)-(192.6) for y.

Thus, we start with the consideration of (192.1)-(192.3).

9.2. Solutions (B,h) of (192.1)-(192.3). A rather simple construction procedure exists in the following important special situation.

9.2.1. <u>The case $\alpha_1=\alpha_2=\ldots=\alpha_r=\frac{1}{r}$.</u> In the present case $\alpha_1=\alpha_2=\ldots=\alpha_r=\frac{1}{r}$, hence $\alpha=\frac{1}{r}1_r$, we still assume that the real random variables

$$\xi_1(\omega),\xi_2(\omega),\ldots,\xi_L(\omega) \text{ are uncorrelated.} \tag{198}$$

Then, in addition to (188), we find that

$$0 = E\ \xi_t(\omega)\xi_\tau(\omega) = \frac{1}{r}\sum_{i=1}^{r}\xi_t^i\,\xi_\tau^i = \frac{1}{r}\,\Xi_t\Xi_\tau' \tag{198a}$$

$$\text{for all } t,\tau=1,\ldots,L,\ t\neq\tau.$$

Since equation (188.2) reads

$$0 = \frac{1}{r}\,\Xi_t\,1_r,\quad t=1,\ldots,L,$$

according to (188.2) and (198a) we have that

$$1_r,\ \Xi_1',\ \Xi_2',\ldots,\Xi_L' \text{ are mutually orthogonal,} \tag{199}$$

and therefore, linear independent r-vectors. Consequently, L, r must fulfill the inequality

$$1+L\leq r,$$

and it is

$$\operatorname{rank}\Xi = L,$$

cf. § 7.1.6.1. Based on (199), solutions (B,h) of (192.1)-(192.3) may be easily found by assuming that

$$B \text{ is a symmetric } r\times r \text{ matrix.} \tag{200}$$

Indeed, if (200) holds true, then (192.1)-(192.3) reads

$$B\,1_r = 1_r$$

$$B\,\Xi_t' = h_{tt}\,\Xi_t',\quad t=1,2,\ldots,L \tag{201.2}$$

$$B \geq 0. \tag{201.3}$$

Thus, first we have to solve L+1 ordinary eigenvalue problems. Since $\{1_r,\Xi_1',\ldots,\Xi_L'\}$ is a set of orthogonal r-vectors, we may further select $r-(L+1)$ orthogonal row vectors $u_{L+2},\ldots,u_r$ in $\mathbb{R}^r$ such that

$$\{1,\Xi_1',\ldots,\Xi_L',\ u_{L+2}',\ldots,u_r'\} \tag{202}$$

is an orthogonal basis of $\mathbb{R}^r$ being composed of eigenvectors of B. Therefore, the solutions B of (201.1)-(201.2) have the form

$$B = \frac{1}{r}\,1_r\,1_r' + \sum_{t=1}^{L} h_{tt}\,\frac{\Xi_t\Xi_t'}{||\Xi_t||^2} + \sum_{t=L+2}^{r} \lambda_t\,\frac{u_t'\,u_t}{||u_t||^2}, \tag{203}$$

where $\lambda_{L+2},\ldots,\lambda_r$ are the eigenvalues of B related to the eigenvectors $u_{L+2}',\ldots,u_r'$ of B. Hence, (201.1)-(201.2) yields

$B = B(h_{11},\ldots,h_{LL},\ \lambda_{L+2},\ldots,\lambda_r,\ u_{L+2},\ldots,u_r)$.

Consequently, we find the following theroem:

Theorem 9.1. Suppose $\alpha = \frac{1}{r}\,1_r$, and let x be a given element of D. Define the matrix B by (203), and select $(y,h) \in \mathbb{R}^n \times \mathbb{R}^L$ such that

$$A^{(0)}y = A^{(0)}x \tag{204.1}$$

$$A^{(t)}y-b^{(t)} = h_{tt}(A^{(t)}x-b^{(t)}),\ t=1,2,\ldots,L \tag{204.2}$$

$$y \in D \tag{204.3}$$

$$B(h_{11},\ldots,h_{LL},\lambda_{L+2},\ldots,\lambda_r,u_{L+2},\ldots,u_r)\geq 0, \tag{204.4}$$

where $\lambda_{L+2},\ldots,\lambda_r$ are any real numbers and $u_{L+2},\ldots,u_r$ are r-vectors such that (202) is an orthogonal basis of $\mathbb{R}^r$, then (y,B,h) is a solution of (192.1)-(192.6).

As a further consequence of the above considerations for H, defined by (197), we find that

$$H \supset H_1 = \{h \in \mathbb{R}^L:\ \text{there are numbers } \lambda_{L+2},\ldots,\lambda_r \text{ and r-vectors } u_{L+2},\ldots,u_r \text{ such that (202) is an othogonal basis of } \mathbb{R}^r \text{ and inequality (204.4) holds}\}. \tag{205}$$

It is easy to see that H_1 contains h = 0.

Remark

a) Since $||B|| \leq (\sum_{i,j=1}^{r} b_{ij}^2)^{1/2}$, relations (201) yield

$$|h_{tt}| \leq \sqrt{r},\ t=1,2,\ldots,L, \tag{206.1}$$

hence $||h||_\infty \leq \sqrt{r}$ for every $h \in H$, where $||\cdot||_\infty$ denotes the maximum-norm. Because of $Bu_t' = \lambda_t u_t'$, $B1_r = 1_r$, and $B \geq 0$, we also find that

$$|\lambda_t| \leq \sqrt{r},\ t = L+2,\ldots,r. \tag{206.2}$$

The bound for the eigenvalues of B given in (206) is not always sharp as is shown by the following example: Let L=1, r=4 and $\Xi_1 = (1,-1,1,-1)$, then (201) yields $|h_{11}| \leq 1$, where $\sqrt{r}=2$.

b) If $\lambda_t=0$ for all $t=L+2,\ldots,r$, then, according to (203), it is

$$B(h) = \frac{1}{r} 1_r 1_r' + \sum_{t=1}^{L} h_{tt} \frac{\Xi_t'\Xi_t}{||\Xi_t||^2}, \tag{203a}$$

and the inequality (204.4) is then reduced to

$$B(h) \geq 0. \tag{204.4a}$$

For h=0 it is $B(h) = \frac{1}{r} 1_r 1_r'$, cf. (193). Moreover, (205) yields

$$H \supset H_1 \supset H_2 = \{h \in \mathbb{R}^L: \frac{1}{r} 1_r 1_r' + \sum_{t=1}^{L} h_{tt} \frac{\Xi_t'\Xi_t}{||\Xi_t||^2} \geq 0\} \tag{207}$$

In the example considered above it is $H_2 = [-\frac{1}{4},\frac{1}{4}]$.

Because of (207) we find that there are infinitely many tuples $(h_{11},\ldots,h_{LL},\lambda_{L+2},\ldots,\lambda_r,u_{L+2},\ldots,u_r)$ such that (202) is an orthogonal basis of $\mathbb{R}^r$ and inequality (204.4) holds true. Indeed, we only have to mention that for every $L \geq 1$, h=0 is an interior point of the closed convex polyhedron H_2.

9.2.2. Arbitrary probability distributions $\alpha=(\alpha_1,\alpha_2,\ldots,\alpha_r)$.

If α is an arbitrary probability distribution, then we must drop the assumption that B is symmetric. Thus, we have to consider the entire system (192.1)-(192.3), i.e.

$$1_r B = 1_r' \tag{208.1}$$

$$B\alpha = \alpha \tag{208.2}$$

$$\Xi_t B = h_{tt}\,\Xi_t,\ t=1,2,\ldots,L \tag{208.3}$$

$$B \geq 0. \tag{208.4}$$

Inserting $B_o = \alpha 1_r'$ into (208), we find that the relations (208.1), (208.2), (208.4) are fulfilled, and (208.3) yields h=0, cf. (188) and (193). Hence, we set

$$B = B_o + C = \alpha 1_r' + C, \tag{209}$$

where C is an rxr matrix which is determined by the following insertion of (209) into (208.1)-(208.3), yielding to

$$1_r' = 1_r'B = 1_r'(\alpha 1_r' + C) = 1_r' + 1_r'C,$$

$$\alpha = B\alpha = (\alpha 1_r' + C)\alpha = \alpha + C\alpha,$$

$$h_{tt}\Xi_t = \Xi_t B = \Xi_t(\alpha 1_r' + C) = \Xi_t C,$$

see (188). Hence, for C we obtain the equations

$$1_r'C = 0 \tag{210.1}$$

$$C\alpha = 0 \tag{210.2}$$

$$\Xi_t C - h_{tt}\Xi_t = 0,\ t=1,2,\ldots,L. \tag{210.3}$$

Since $h=(h_{11},h_{22},\ldots,h_{LL})'$ is a vector of free parameters, (210) represents a homogeneous linear system of 2r+Lr equations for the tuple (C,h) containing r^2+L unknowns. Let

$$W_o = \{(C,h) : (C,h) \text{ solves } (210)\} \tag{211}$$

denote the linear space of solutions (C,h) of (210). We find that

$$\dim \mathcal{W}_o = r^2+L - \text{rank}(210) \geq r^2+L-r(2+L).$$

Obviously, if $r\geq 2+L$, then $\dim \mathcal{W}_o \geq L$.

According to (209)-(211), the set $\mathcal{W}$ of solutions (B,h) of (208) has the form

$$\mathcal{W} = \{(\alpha 1_r' + C,h): (C,h)\in \mathcal{W}_o,\ \alpha 1_r'+C\geq 0\}. \tag{212}$$

Thus, for the set H defined in (197), we find

$$H = \{h\in \mathbb{R}^L: \text{there is a matrix } C \text{ such that } (C,h)\in \mathcal{W}_o \text{ and } \alpha 1_r' + C \geq 0\}.$$

Some properties of $\mathcal{W}$ are given in the next lemma:

Lemma 9.2. a) $\mathcal{W}$ is a closed convex subset of $\mathbb{R}^{r\cdot r}\times \mathbb{R}^L$ and $\mathcal{W}$ contains $(\alpha 1_r',0)$. b) If $\dim \mathcal{W}_o>0$, then $\mathcal{W}$ contains a certain line segment through $(\alpha 1_r',0)$. c) If $(C,h)\in \mathcal{W}_o$ with $h\neq 0$, then there is a number $t_o>0$ such that $(\alpha 1_r'+tC,th)\in \mathcal{W}$, where $th\neq 0$ for all $|t|\leq t_o$.

Proof. The first assertion follows from (212) and (193). If $\dim \mathcal{W}_o>0$, then $\mathcal{W}_o$ contains an element $(C,h)\neq 0$. Hence, $(tC,th)\in \mathcal{W}_o$ for all $t\in \mathbb{R}$, and there is a number $t_o>0$ such that $\alpha 1_r'+tC\geq 0$ for all $|t|\leq t_o$. Thus, $(\alpha 1_r'+tC,th)\in \mathcal{W}$ for all $|t|\leq t_o$. The last assertion can be shown as before.

Now, we are going to determine the space $\mathcal{W}_o$ of solutions (C,h) of (210). Let c_{ij}, $1\leq i,j\leq r$, denote the elements of C. The equations (210.1) and (210.2) yield to

$$c_{rj} = -\sum_{i=1}^{r-1} c_{ij},\ j=1,2,\dots,r \tag{213.1}$$

$$c_{ir} = -\sum_{j=1}^{r-1} \frac{\alpha_j}{\alpha_r} c_{ij},\ i=1,2,\dots,r. \tag{213.2}$$

Consequently, for i=j=r we get the condition

$$c_{rr} = -\sum_{i=1}^{r-1} c_{ir} = -\sum_{j=1}^{r-1} \frac{\alpha_j}{\alpha_r} c_{rj}. \tag{214}$$

Using representation (213) of c_{ir}, $1\leq i\leq r-1$, and c_{rj}, $1\leq j\leq r-1$, in (214), we find

$$-\sum_{i=1}^{r-1} \left(-\sum_{j=1}^{r-1} \frac{\alpha_j}{\alpha_r} c_{ij}\right) = -\sum_{j=1}^{r-1} \frac{\alpha_j}{\alpha_r} \left(-\sum_{i=1}^{r-1} c_{ij}\right).$$

Hence, the second equation in (214) holds identically for all c_{ij}, $1\leq i,j\leq r-1$. Therefore, the solutions $C = (c_{ij})$ of (210.1) and (210.2) are given by

$$c_{rj} = -\sum_{i=1}^{r-1} c_{ij}, \; j=1,2,\ldots,r-1 \tag{215.1}$$

$$c_{ir} = -\sum_{j=1}^{r-1} \frac{\alpha_j}{\alpha_r} c_{ij}, \; i=1,2,\ldots,r-1 \tag{215.2}$$

$$c_{rr} = \sum_{i=1}^{r-1}\sum_{j=1}^{r-1} \frac{\alpha_j}{\alpha_r} c_{ij}, \tag{215.3}$$

where the elements c_{ij}, $1\leq i,j\leq r-1$, can take arbitrary values.

In order to handle (210.3), we first compute $\Xi_t C$ for every $1\leq t\leq L$. The j-th component of $\Xi_t C$ is given by $\sum_{i=1}^{r} \xi_t^i c_{ij}$. Hence, using (215.1), for $1\leq j\leq r-1$ we find

$$\sum_{i=1}^{r} \xi_t^i c_{ij} = \sum_{i=1}^{r-1} \xi_t^i c_{ij} + \xi_t^r c_{rj} = \sum_{i=1}^{r-1} \xi_t^i c_{ij} + \xi_t^r\left(-\sum_{i=1}^{r-1} c_{ij}\right) =$$

$$= \sum_{i=1}^{r-1} (\xi_t^i - \xi_t^r) c_{ij}.$$

Consequently, according to (210.3), for every t=1,2,...,L we have the conditions

$$\sum_{i=1}^{r-1} (\xi_t^i - \xi_t^r) c_{ij} = h_{tt}\, \xi_t^j, \; j=1,2,\ldots,r-1. \tag{216}$$

We now consider the case j=r. By (215.2), (215.3) and (216) it is

$$\sum_{i=1}^{r} \xi_t^i c_{ir} = \sum_{i=1}^{r-1} \xi_t^i c_{ir} + \xi_t^r c_{rr} =$$

$$= \sum_{i=1}^{r-1} \xi_t^i \left(- \sum_{j=1}^{r-1} \frac{\alpha_j}{\alpha_r} c_{ij}\right) + \xi_t^r \left(\sum_{i=1}^{r-1} \sum_{j=1}^{r-1} \frac{\alpha_j}{\alpha_r} c_{ij}\right) =$$

$$= \sum_{j=1}^{r-1} \frac{\alpha_j}{\alpha_r} \sum_{i=1}^{r-1} (- \xi_t^i + \xi_t^r) c_{ij} = \sum_{j=1}^{r-1} \frac{\alpha_j}{\alpha_r} (-h_{tt} \xi_t^j) =$$

$$= -\frac{h_{tt}}{\alpha_r} \sum_{j=1}^{r-1} \xi_t^j \alpha_j = - \frac{h_{tt}}{\alpha_r} (-\xi_t^r \alpha_r) = h_{tt} \xi_t^r,$$

since $0 = \Xi_t \alpha = \sum_{j=1}^{r} \xi_t^j \alpha_j$. Thus, the r-th equation of (210.3) is implied by (216).

Denoting by $\tilde{C}$ the $(r-1)x(r-1)$ submatrix $\tilde{C} = (c_{ij})_{i,j=1,..,r-1}$ of C, we now find the next theorem:

Theorem 9.2. a) Using representation (215) of c_{rj}, c_{ir}, $1 \leq i,j \leq r-1$, and c_{rr} as functions of $\tilde{C}$, system (210) for (C,h) is equivalent to the homogeneous linear system

$$\begin{pmatrix} \xi_1^1 - \xi_1^r & \xi_1^2 - \xi_1^r & \dots & \xi_1^{r-1} - \xi_1^r \\ \xi_2^1 - \xi_2^r & \xi_2^2 - \xi_2^r & \dots & \xi_2^{r-1} - \xi_2^r \\ \vdots & \vdots & & \vdots \\ \xi_L^1 - \xi_L^r & \xi_L^2 - \xi_L^r & \dots & \xi_L^{r-1} - \xi_L^r \end{pmatrix} \begin{pmatrix} c_{1j} \\ c_{2j} \\ \vdots \\ c_{r-1j} \end{pmatrix} - \begin{pmatrix} h_{11} \xi_1^j \\ h_{22} \xi_2^j \\ \vdots \\ h_{LL} \xi_L^j \end{pmatrix} = 0, \quad (217)$$

$$j=1,2,\dots,r-1,$$

for the tuple $(\tilde{C},h)$. b) If $\tilde{W}_o$ denotes the linear space of solutions $(\tilde{C},h)$ of (217), then

$$\dim \tilde{W}_o \geq (r-1)^2 + L - (r-1)L.$$

c) For given $h \in \mathbb{R}^L$, the columns $\tilde{c}_j$, $1 \leq j \leq r-1$, of $\tilde{C}$ are determined by r-1 systems of linear equations having an identical system matrix

$$\Psi = \begin{pmatrix} \xi_1^1 - \xi_1^r & \xi_1^2 - \xi_1^r & \dots & \xi_1^{r-1} - \xi_1^r \\ \xi_2^1 - \xi_2^r & \xi_2^2 - \xi_2^r & \dots & \xi_2^{r-1} - \xi_2^r \\ \vdots & \vdots & & \vdots \\ \xi_L^1 - \xi_L^r & \xi_L^2 - \xi_L^r & \dots & \xi_L^{r-1} - \xi_L^r \end{pmatrix}. \tag{218}$$

d) It is rank $\Psi = L$ if and only if

$$1_r, \Xi_1', \Xi_2', \dots \Xi_L' \text{ are linear independent} \tag{219}$$

r- vectors.

Proof. We still have to show the last part of the theorem. Adding to Ψ an r-th zero column, it is easy to see that rank $\Psi = L$ if and only if the row vectors

$$v_t = \Xi_t - \xi_t^r \, 1_r', \; t=1,2,\dots,L$$

are linear independent. Since

$$\sum_{t=1}^{L} \mu_t v_t = (-\sum_{t=1}^{L} \mu_t \xi_t^r) 1_r' + \sum_{t=1}^{L} \mu_t \Xi_t,$$

we see that (219) yields the linear independence of $v_1,\dots,v_L$. Conversely, suppose that $v_1,\dots,v_L$ are linear independent. If (219) does not hold, then there are numbers $\mu_o,\mu_1,\dots,\mu_L$ such that $|\mu_o| + \sum_{t=1}^{L} |\mu_t| > 0$ and

$$\mu_o \, 1_r' + \sum_{t=1}^{L} \mu_t \Xi_t = 0.$$

This yields $\sum_{t=1}^{L} |\mu_t| > 0$ and $\mu_o + \sum_{t=1}^{L} \mu_t \xi_t^r = 0$. Hence, we obtain

$$\sum_{t=1}^{L} \mu_t v_t = \sum_{t=1}^{L} \mu_t (\Xi_t - \xi_t^r \, 1_r') = \sum_{t=1}^{L} \mu_t \Xi_t + (-\sum_{t=1}^{L} \mu_t \xi_t^r) 1_r' =$$

$$= \mu_o \, 1_r' + \sum_{t=1}^{L} \mu_t \Xi_t = 0,$$

where $\sum_{t=1}^{L} |\mu_t|>0$. This is a contradiction to the assumption that $v_1,\ldots,v_L$ are linear independent. Hence, (219) must hold.

Remark

Under the assumptions in § 9.2.1, condition (219) is fulfilled, c.f. (199). Furthermore, the assumptions in § 9.2.1 yield that $h=0$ is an interior point of H, cf. (207). A corresponding result can be proved also in the general case.

Theorem 9.3. If (219) holds, then $h=0$ is an interior point of H.

Proof. If (219) holds, then, by Theorem 9.2, we know that rank $\Psi =$ $=L$. Consequently, according to (217) and (215), we find that for every given $h \in \mathbb{R}^L$ our system (210) has a solution $C = (c_1,c_2,\ldots,c_r)$ of the type

$$c_j = K_j h, \; j=1,2,\ldots,r,$$

where K_j is a certain rxL matrix. Thus, we have a positive number ρ_o such that

$$B(h) = \alpha 1_r' + C = \alpha 1_r' + (K_1 h, K_2 h, \ldots, K_r h) \geq 0$$

for every $h \in \mathbb{R}^L$ such that $||h|| \leq \rho_o$. Hence, $(B(h),h)$ is a solution of (208) for $||h|| \leq \rho_o$. This means that $h \in H$ for every $h \in \mathbb{R}^L$ such that $||h|| \leq \rho_o$.

9.3. Solutions of (192.4)-(192.6) in case $D = \mathbb{R}^n$. Considering now the remaining conditions (192.4)-(192.6), for $D = \mathbb{R}^n$ we have to solve the following system of, at the most, $(L+1)m$ linear equations

$$A^{(0)}y = A^{(0)}x \tag{220.1}$$

$$A^{(t)}y = b^{(t)} + h_{tt}(A^{(t)}y-b^{(t)}), \; t=1,\ldots,L \tag{220.2}$$

for $y \in \mathbb{R}^n$, where $h=(h_{11},\ldots,h_{LL})'$ is a given element of $H \setminus \{1_L\}$, cf. (197). We know that h=0 lies in H, furthermore, if the vectors $1_L, \Xi_1', \Xi_2', \ldots, \Xi_L'$ are linear independent, then h=0 is also an interior point of H, cf. (207) and Theorem 9.3. For h=0 system (220) is reduced to (195), i.e.

$$\begin{aligned} A^{(0)}y &= A^{(0)}x \\ A^{(t)}y &= b^{(t)},\ t=1,\ldots,L. \end{aligned} \tag{195}$$

According to Lemma 9.1 we know that

$$S_o \subset S_{\mathbb{R}^n}, \tag{221.1}$$

where S_o is defined by

$$S_o = \{x \in \mathbb{R}^n\colon A^{(t)}x = b^{(t)},\ t=1,\ldots,L\}, \tag{221.2}$$

and $S_{\mathbb{R}^n}$ denotes the set of $\mathbb{R}^n$-stationary points.
Moreover, if for every $x \in \mathbb{R}^n$ (195) has a solution y, then $S_{\mathbb{R}^n}=S_o$. Obviously, every solution of (195) is an $\mathbb{R}^n$-stationary point. If $x \in S_o$, then the solutions y of (220) are again elements of S_o for arbitrary $h \in \mathbb{R}^L$.

Interpreting $h \in \mathbb{R}^L$ as a variable to be taken from $H \setminus \{1_L\}$, (220) can also be considered as a system of, at the most, (L+1)m linear equations

$$\begin{aligned} A^{(0)}y &= A^{(0)}x \\ A^{(t)}y-h_{tt}(A^{(t)}x-b^{(t)}) &= b^{(t)},\ t=1,\ldots,L \end{aligned} \tag{222.1}$$

for the (n+L)-tuple $(y,h) \in \mathbb{R}^n \times H$. The above system always has the trivial feasible solution $(y,h)=(x,1_L)$.
Consequently, setting $(y,h)=(x,1_L)+(x,\eta)$, we have to find solutions (w,η) of the homogeneous linear system related to (222.1)

$$\begin{aligned} A^{(0)}w &= 0 \\ A^{(t)}w-\eta_{tt}(A^{(t)}x-b^{(t)}) &= 0,\ t=1,\ldots,L \end{aligned} \tag{222.2}$$

such that

$$h = 1_L + \eta \in H \setminus \{1_L\}.$$

Having to solve (220), (222), we must also determine, of course, the following set

$$V = \{x \in \mathbb{R}^n: \text{ there is } h \in H \text{ with } h_{tt} \neq 1 \text{ for all } t=1,\ldots,L \text{ such that (220) has a solution } y \in \mathbb{R}^n\}. \tag{223}$$

If $x \in V \setminus S_o$, then by (221.2), (223) and the definition (197) of H, there is a tuple (y,B,h) solving (192) such that $A^{(t)}y \neq A^{(t)}x$ for at least one $1 \leq t \leq L$. Hence, by Definition 4.1, x is not $\mathbb{R}^n$-stationary. Thus, we have proven this lemma:

Lemma 9.3. Let V be defined by (223). Then

$$S_{\mathbb{R}^n} \subset S_o \cup V^c,$$

where $V^c = \mathbb{R}^n \setminus V$.

Note. If $V = \mathbb{R}^n$, then (221.1) and Lemma 9.3 yield $S_{\mathbb{R}^n} = S_o$.

For the following discussion of (220), we first note that (220.1) can be replaced by

$$\tilde{A}^{(0)}y = \tilde{A}^{(0)}x, \tag{220.1a}$$

where $\tilde{A}^{(0)}$ is a $\rho^{(0)} \times n$ submatrix of $A^{(0)}$ such that

$$\operatorname{rank} \tilde{A}^{(0)} = \operatorname{rank} A^{(0)} = \rho^{(0)}.$$

Define now the $(\rho^{(0)} + Lm) \times n$ matrix

$$\mathbb{A} = \begin{pmatrix} \tilde{A}^{(0)} \\ A^{(1)} \\ \cdot \\ \cdot \\ \cdot \\ A^{(L)} \end{pmatrix}, \tag{224}$$

see matrix $\hat{\mathbb{A}}$ in § 7.1. Furthermore, let $\rho = \operatorname{rank} \mathbb{A}$, hence

$\rho \leq n$ and $\rho \leq \rho^{(0)}+Lm$. By means of $\mathbb{A}$ (220) can be represented by

$$\mathbb{A}y = \begin{pmatrix} \tilde{A}^{(0)}x \\ b^{(1)}+h_{11}(A^{(1)}x-b^{(1)}) \\ \cdot \\ \cdot \\ \cdot \\ b^{(L)}+h_{LL}(A^{(L)}x-b^{(L)}) \end{pmatrix}. \tag{225}$$

Since the random matrix $(A(\omega),b(\omega))$ is represented here by (60), there is a close relationship between (225) and system (46.3a) in § 7.1.

We now have to discuss two cases:

9.3.1. <u>Rank $\mathbb{A}$ = rank $A^{(0)}$ + Lm.</u> We suppose next to that the rows of the matrices $\tilde{A}^{(0)},A^{(1)},\ldots,A^{(L)}$ are linear independent. Obviouly, this may occur only under the condition $\rho^{(0)}+Lm \leq n$, hence

$$L \leq \frac{1}{m}\,(n - \text{rank } A^{(0)}). \tag{226}$$

Note that (226) is a condition for the maximal number of random elements $\xi_t(\omega)$ in $(A(\omega),b(\omega))$. Since $0 \leq \rho^{(0)} = \text{rank } A^{(0)} \leq \min\{m,n\}$, and in most practical cases it is $m<n$ or even $m<<n$, for the right hand side of (226) we usually have

$$\frac{1}{m}(n\text{-rank } A^{(0)}) \geq \frac{1}{m}(n-m) = \frac{n}{m} - 1 > (>>)0.$$

Using representation (225) of (220), we find now this

<u>Theorem 9.4.</u> Let $D=\mathbb{R}^n$, and suppose that rank $\mathbb{A}$ = rank $A^{(0)}+Lm$, where $\mathbb{A}$ is given by (224), then $V = \mathbb{R}^n$ and

$$S_{\mathbb{R}^n} = S_o = \{x \in \mathbb{R}^n\colon A^{(t)}x=b^{(t)},\ t=1,\ldots,L\},$$

where S_o is an (n-Lm)-dimensional linear manifold in $\mathbb{R}^n$.

Proof. If rank $\mathbb{A}$ is equal to the number of rows of $\mathbb{A}$, then (225) has solutions y for arbitrary vectors $x \in \mathbb{R}^n$ and $h \in \mathbb{R}^L$, hence $\mathcal{V}=\mathbb{R}^n$, see (223). Moreover, since (195) is contained in (225) for h=0, the rest of the theorem follows now from Lemma 9.1.

We consider now the second main case:

9.3.2. <u>Rank $\mathbb{A}$ < rank $A^{(0)}$ + Lm.</u> Using a similar procedure as in the discussion of equation (46.3a) in part IV of section 7.1.3, we select here a $\rho\times n$ submatrix

$$\mathbb{A}_I = (A_i^{(t)})_{(t,i)\in K} \tag{227.1}$$

of $\mathbb{A}$ such that rank $\mathbb{A}_I=\rho$, where K is an index set containing ρ pairs (t,i), $0\le t\le L$, $1\le i\le m$, and $A_i^{(t)}$ denotes the i-th row of $A^{(t)}$. Defining $h_{oo}=1$, $b^{(0)}=0$, and

$$\mathbb{A}_{II} = (A_i^{(t)})_{(t,i)\notin K},$$

equation (225) reads

$$\mathbb{A}_I y = (b_i^{(t)}+h_{tt}(A_i^{(t)}x-b_i^{(t)}))_{(t,i)\in K}, \tag{228.1}$$

$$\mathbb{A}_{II} y = (b_i^{(t)}+h_{tt}(A_i^{(t)}x-b_i^{(t)})_{(t,i)\notin K}. \tag{228.2}$$

Since rank $\mathbb{A}_I=\rho$, $\mathbb{A}_I$ can be decomposed into a regular $\rho\times\rho$ submatrix $\mathbb{A}_{I,I}$ and into a complementary $\rho\times(n-\rho)$ submatrix $\mathbb{A}_{I,II}$. Decomposing then y in the same way into subvectors $y_I\in\mathbb{R}^\rho$ and $y_{II}\in\mathbb{R}^{n-\rho}$, (228.1) yields

$$y_I = \mathbb{A}_{I,I}^{-1}(b_i^{(t)}+h_{tt}(A_i^{(t)}x-b_i^{(t)}))_{(t,i)\in K} - \mathbb{A}_{I,I}^{-1}\,\mathbb{A}_{I,II}p, \tag{229}$$

where $y_{II}=p\in\mathbb{R}^{n-\rho}$ may be chosen arbitrarily. Inserting (229) into (228.2), system (225) can be represented by

$$\begin{aligned}(b_i^{(t)} + h_{tt}(A_i^{(t)}x-b_i^{(t)}))_{(t,i)\notin K} &= \\ = \mathbb{A}_{II,I}\,\mathbb{A}_{I,I}^{-1}(b_i^{(t)}+h_{tt}(A_i^{(t)}x-b_i^{(t)}))_{(t,i)\in K} \\ - \mathbb{A}_{II,I}\,\mathbb{A}_{I,I}^{-1}\,\mathbb{A}_{I,II}p + \mathbb{A}_{II,II}p,\end{aligned} \tag{230}$$

where $\mathbb{A}_{II,I}$, $\mathbb{A}_{II,II}$ is the decomposition of $\mathbb{A}_{II}$ corresponding to the decomposition of $\mathbb{A}_I$ into $\mathbb{A}_{I,I}$ and $\mathbb{A}_{I,II}$. For arbitrary given vectors $h\in\mathcal{H}$ and $p\in\mathbb{R}^{n-\rho}$ equation (230) is a condition for x.

Thus, we have this result:

Theorem 9.5. Let $\rho<\rho^{(0)}+Lm$, then V is given by

$$V = \{x \in \mathbb{R}^n: \Psi(h)x = \mathbb{A}_{II,I}\,\mathbb{A}^{-1}_{I,I}((1-h_{tt})b_i^{(t)})_{(t,i)\in K} - ((1-h_{tt})b_i^{(t)})_{(t,i)\notin K} + (\mathbb{A}_{II,II}-\mathbb{A}_{II,I}\,\mathbb{A}^{-1}_{I,I}\,\mathbb{A}_{I,II})p:\ h\in H,\ h_{tt}\neq 1,\ 1\leq t\leq L,\ h_{oo}=1,\ p\in\mathbb{R}^{n-\rho}\}, \tag{231.1}$$

where the $(\rho^{(0)}+Lm-\rho)\times n$ matrix $\Psi(h)$ is defined by

$$\Psi(h) = (h_{tt}A_i^{(t)})_{(t,i)\notin K} - \mathbb{A}_{II,I}\,\mathbb{A}^{-1}_{I,I}(h_{tt}\,A_i^{(t)})_{(t,i)\in K}. \tag{231.2}$$

Note

Equation (231.1) is related to conditions (52), (52.a) in § 7.1.3.IV. Simplifications of (231.1) arise under special assumptions, e.g. if $A_{II,I}=0$, $A_{II,II}=0$, $A_{I,II}=0$, respectively. Moreover, in the important special case

$$b^{(t)}=0,\ t=1,\dots,L, \tag{232}$$

(231.1) is reduced ($b^{(0)}=0$) to

$$\Psi(h)x = (\mathbb{A}_{II,II}-\mathbb{A}_{II,I}\,\mathbb{A}^{-1}_{I,I}\,\mathbb{A}_{I,II})p. \tag{233}$$

If (232) holds true, and $\rho=\text{rank}\,\mathbb{A}_I=n$, then $\mathbb{A}_{I,I}=\mathbb{A}_I$, $\mathbb{A}_{II,I}=\mathbb{A}_{II}$, and (233) reads

$$\Psi(h)x = 0.$$

Hence, in this case, V is a cone.

9.4. Solutions of (192.4) and (192.5) in special situations.

Suppose that $S_o\neq 0$, and select a vector

$$\xi\in S_o. \tag{234}$$

Having (234) and defining $h_{oo}=1$, then, because of $b^{(t)}=A^{(t)}\xi$, $t=1,\dots,L$, system (220) obviously reads

$$A^{(t)}(y-\xi) = h_{tt}A^{(t)}(x-\xi),\ t=0,1,\dots,L. \tag{235}$$

We observe that, for each t=0,1,...,L,

$$y^{(t)} = \xi + h_{tt}(x-\xi)$$

is a solution of the t-th subequation of (235). Furthermore, for $A^{(0)}=0$ we immediately get the following theorem:

Theorem 9.6. Let $D = \mathbb{R}^n$, and suppose that $S_o \neq \emptyset$, $A^{(0)}=0$, then $S_{\mathbb{R}^n}=S_o$ and $V=\mathbb{R}^n$. Moreover, for every $x \in \mathbb{R}^n$ and $\lambda \in \mathbb{R}$,

$$y = \xi + \lambda(x-\xi) \tag{236}$$

is a solution of (220), where ξ is an arbitrary, but fixed element of S_o.

Proof. If $S_o \neq \emptyset$ and $A^{(0)}=0$, then Lemma 9.1 yields $S_{\mathbb{R}^n}=S_o$. Putting $h_{tt}=\lambda$, t=1,...,L, we find that the vector y given by (236) solves (235), and therefore also (220), for every $x \in \mathbb{R}^n$ and $\lambda \in \mathbb{R}$. Consequently, for every tuple $(x,h=0)$, $x \in \mathbb{R}^n$, system (220) has the solution $y=\xi$. Hence, it is $V=\mathbb{R}^n$.

If $\bar{A} = A^{(0)} \neq 0$, then we have to study (235) in more detail. For this purpose, let $\rho^{(t)} = \text{rank } A^{(t)} = \text{rank }(A^{(t)},b^{(t)})$, t=0,1,...,L, and select a $\rho^{(t)} \times n$ submatrix $\tilde{A}^{(t)}$ of $A^{(t)}$ such that $\text{rank } \tilde{A}^{(t)} = \rho^{(t)}$, t=0,1,...,L. Corresponding to (224), here we then have to consider the $(\sum_{t=0}^{L} \rho^{(t)}) \times n$ matrix

$$\tilde{\mathbb{A}} = \begin{pmatrix} \tilde{A}^{(0)} \\ \tilde{A}^{(1)} \\ \cdot \\ \cdot \\ \cdot \\ \tilde{A}^{(L)} \end{pmatrix} . \tag{237}$$

Let $\tilde{\rho} = \text{rank } \tilde{\mathbb{A}}$. By means of $\tilde{\mathbb{A}}$ we may represent (235), and therefore also (220), by

$$\tilde{\mathbb{A}}(x-\xi) = \begin{pmatrix} \tilde{A}^{(0)}(x-\xi) \\ h_{11}A^{(1)}(x-\xi) \\ \vdots \\ h_{LL}A^{(L)}(x-\xi) \end{pmatrix} \tag{238}$$

which is a system of $\sum_{t=0}^{L} \rho^{(t)}$ linear equations for $y-\xi=\hat{y} \in \mathbb{R}^n$. Obviously, (238) can be handled as system (225) in sections 9.3.1 and 9.3.2.

9.4.1. Rank $\tilde{\mathbb{A}} = \sum_{t=0}^{L} \text{rank } A^{(t)}$. Under this assumption we must have

$$\sum_{t=0}^{L} \rho^{(t)} \leq n$$

which is often a much weaker condition than (226). Since (195) has here the form

$$\tilde{\mathbb{A}}y = \begin{pmatrix} \tilde{A}^{(0)}x \\ \tilde{A}^{(1)}\xi \\ \cdot \\ \cdot \\ \cdot \\ \tilde{A}^{(L)}\xi \end{pmatrix},$$

we reach the following theorem:

Theorem 9.7. Let $D=\mathbb{R}^n$, $A^{(0)} \neq 0$, and suppose that (234) holds true. If $\tilde{\rho} = \sum_{t=0}^{L} \rho^{(t)}$, then $V=\mathbb{R}^n$ and

$$S_{\mathbb{R}^n} = S_o = \xi + \{x \in \mathbb{R}^n : \tilde{\mathbb{A}}^{(1)} = 0\},$$

where $\dim(S_o-\xi) = n - \sum_{t=1}^{L} \rho^{(t)}$ and $\tilde{\mathbb{A}}^{(1)}$ is given by

$$\tilde{\mathbb{A}}^{(1)} = \begin{pmatrix} \tilde{A}^{(1)} \\ \cdot \\ \cdot \\ \cdot \\ \tilde{A}^{(L)} \end{pmatrix}.$$

Proof. Having the equivalence between (238) and (220), the above assertion can be proven by using the same arguments as in the proof of Theorem 9.4.

9.4.2. Rank $\tilde{\mathbb{A}} < \sum_{t=0}^{L}$ rank $A^{(t)}$. Corresponding to section 9.3.2, we have here a $\tilde{\rho} \times n$ submatrix

$$\tilde{\mathbb{A}}_I = (A_i^{(t)})_{(t,i)\in\tilde{K}} \tag{239.1}$$

of $\tilde{\mathbb{A}}$ such that rank $\tilde{\mathbb{A}}_I = \tilde{\rho}$, where $A_i^{(t)}$ is the i-th row of $A^{(t)}$. If

$$\tilde{\mathbb{A}}_{II} = (A_i^{(t)})_{(t,i)\notin K} \tag{239.2}$$

denotes the complementary submatrix to $\tilde{\mathbb{A}}_I$, then (238) reads

$$\tilde{\mathbb{A}}_I(y-\xi) = (h_{tt}\ A_i^{(t)}(x-\xi))_{(t,i)\in\tilde{K}} \tag{240.1}$$

$$\tilde{\mathbb{A}}_{II}(y-\xi) = (h_{tt}\ A_i^{(t)}(x-\xi))_{(t,i)\notin\tilde{K}}, \tag{240.2}$$

where $h_{oo}=1$. Decomposing then $\tilde{\mathbb{A}}_I$ into a regular $\tilde{\rho} \times \tilde{\rho}$ submatrix $\tilde{\mathbb{A}}_{I,I}$ and a complementary submatrix $\tilde{\mathbb{A}}_{I,II}$ of $\tilde{\mathbb{A}}_I$, according to (240.1), we obtain

$$(y-\xi)_I = \tilde{\mathbb{A}}_{I,I}^{-1}\ (h_{tt}\ A_i^{(t)}(x-\xi))_{(t,i)\in\tilde{K}} - \tilde{\mathbb{A}}_{I,I}^{-1}\ \tilde{\mathbb{A}}_{I,II}\ \tilde{p}, \tag{241}$$

where $y-\xi$ is decomposed according to $\tilde{\mathbb{A}}_{I,I}$, $\tilde{\mathbb{A}}_{I,II}$ into subvectors $(y-\xi)_I$ and $(y-\xi)_{II} = \tilde{p} \in \mathbb{R}^{n-\tilde{\rho}}$. Decomposing also $\tilde{\mathbb{A}}_{II}$ into submatrices $\tilde{\mathbb{A}}_{II,I}$, $\tilde{\mathbb{A}}_{II,II}$, and putting then (241) into (240.2), corresponding to (230), in the present case, for x we find the condition

$$\tilde{\Psi}(h)(x-\xi) = (\tilde{\mathbb{A}}_{II,II} - \tilde{\mathbb{A}}_{II,I}\ \tilde{\mathbb{A}}_{I,I}^{-1}\ \tilde{\mathbb{A}}_{I,II})\tilde{p}, \tag{242.1}$$

where the $(\sum_{t=0}^{L} \rho^{(t)} - \tilde{\rho}) \times n$ matrix $\tilde{\Psi}(h)$ is defined by

$$\Psi(h) = (h_{tt}\ A_i^{(t)})_{(t,i)\notin\tilde{K}} - \tilde{\mathbb{A}}_{II,I}\ \tilde{\mathbb{A}}_{I,I}^{-1}(h_{tt}\ A_i^{(t)})_{(t,i)\in\tilde{K}} \tag{242.2}$$

and $\tilde{p}$ is an arbitrary $(n-\tilde{\rho})$-vector, cf. (231). Hence, corresponding to Theorem 9.5, we find the following result:

Theorem 9.8. Let $D=\mathbb{R}^n$, and suppose that (234) holds true.

If $\tilde{\rho} < \sum_{t=0}^{L} \rho^{(t)}$, then V is given by

$$V = \{x \in \mathbb{R}^n: \tilde{\Psi}(h)(x-\xi) = (\tilde{\mathbb{A}}_{II,II} - \tilde{\mathbb{A}}_{II,I}\tilde{\mathbb{A}}_{I,I}^{-1}\tilde{\mathbb{A}}_{I,II})\tilde{p} \qquad (243)$$

$$h_{tt} \neq 1,\ 1 \leq t \leq L, h_{00}=1,\ \tilde{p} \in \mathbb{R}^{n-\tilde{\rho}}\}.$$

Note. If $\tilde{\rho}=n$, then (242.1) is reduced to

$\tilde{\Psi}(h)(x-\xi) = 0.$

Hence, V is then a cone with vertex ξ.

9.5. Kernel-representation of solutions y of (220)

In the following we assume that

$$\text{rank}\ A^{(t)} = \rho^{(t)} = \text{rank}(A^{(t)}, b^{(t)}),\ t=1,\ldots,L. \qquad (244)$$

This yields that for every $t=1,\ldots,L$ there is a vector $\xi^{(t)} \in \mathbb{R}^n$ such that

$A^{(t)}\xi^{(t)} = b^{(t)},\ t=1,\ldots,L,$

cf. (234). Putting $h_{oo}=1$ and $b^{(0)}=A^{(0)}\xi^{(0)}$, where $\xi^{(0)}$ is an arbitrary, but fixed n-vector, our basic system (220) reads

$$A^{(t)}(y-\xi^{(t)}-h_{tt}(x-\xi^{(t)})) = 0, t=0,1,\ldots,L. \qquad (245)$$

Hence, we find

$$y-\xi^{(t)}-h_{tt}(x-\xi^{(t)}) = k^{(t)},\ t=0,1,\ldots,L, \qquad (246.1)$$

$$k^{(t)} \in \text{kern}A^{(t)}, \qquad (246.2)$$

where $\text{kern}A^{(t)}$ denotes the kernel of $A^{(t)}$. It is

$$\text{kern}A^{(t)} = \{C^{(t)}p^{(t)}: p^{(t)} \in \mathbb{R}^{n-\rho^{(t)}}\}, \qquad (246.3)$$

where $C^{(t)}$ is an $n \times (n-\rho^{(t)})$ matrix having rank $C^{(t)}=n-\rho^{(t)}$.

For simplicity, we first consider the case L=1.

9.5.1. The case L=1. In the present case it is $\xi^{(1)} \in S_o \subset S_{\mathbb{R}^n}$, and (246) reads

$$y = x + k^{(0)},\ k^{(0)} \in \text{kern}A^{(0)} \qquad (247.1$$

$$y = \xi^{(1)} + h_{11}(x-\xi^{(1)}) + k^{(1)},\ k^{(1)} \in \text{kern}A^{(1)}. \qquad (247.2)$$

Since the case $A^{(0)}=0$ can be completely handled by means of Theorem 9.6, we may suppose that $A^{(0)}\neq 0$. Thus, (247) has a solution y if and only if x and $h_{11}\neq 1$ are related by

$$x = \xi^{(1)} - \frac{1}{1-h_{11}} k^{(0)} + \frac{1}{1-h_{11}} k^{(1)}. \tag{248.1}$$

For every x given by (248.1) system (247) has then the solution

$$y = \xi^{(1)} - \frac{h_{11}}{1-h_{11}} k^{(0)} + \frac{1}{1-h_{11}} k^{(1)}. \tag{248.2}$$

Since

$$A^{(1)}y-b^{(1)} = - \frac{h_{11}}{1-h_{11}} A^{(1)}k^{(0)} = h_{11}(A^{(1)}x-b^{(1)}) \tag{248.3}$$

and $A^{(0)}x = A^{(0)}\xi^{(1)} + \frac{1}{1-h_{11}} A^{(0)}k^{(1)} = A^{(0)}y$, we get this result:

Theorem 9.9. Assume that $D=\mathbb{R}^n$, $L=1$ and $A^{(0)}\neq 0$, then $S_o\neq\xi^{(1)} + \mathrm{kern}A^{(1)}$, and V is given by

$$V = \mathrm{kern}A^{(0)} + S_o. \tag{249.1}$$

If $x = \xi^{(1)} + u^{(0)} + u^{(1)}$, $u^{(t)}\in \mathrm{kern}A^{(t)}$, $t=0,1$, lies in V, then

$$y = \xi^{(1)} + h_{11}u^{(0)} + u^{(1)} \tag{249.2}$$

is a solution of (220) for every $h_{11}\in\mathbb{R}$.

Proof. If $x\in V$, see (223), then (247) and (248) yield that x has the form $x=\xi^{(1)}+u^{(0)}+u^{(1)}$ with some vectors $u^{(t)}\in \mathrm{kern}A^{(t)}$, $t=0,1$. Conversely, suppose that x can be represented as above, and let then y be defined by (249.2). Then $A^{(0)}x=A^{(0)}\xi^{(1)}+A^{(0)}u^{(1)}=A^{(0)}y$ and $A^{(1)}y-b^{(1)}=h_{11}A^{(1)}u^{(0)}=h_{11}(A^{(1)}x-b^{(1)})$ for every $h_{11}\in\mathbb{R}$. Hence $x\in V$.

Remark

a) If $\mathrm{kern}A^{(0)}\subset \mathrm{kern}A^{(1)}$, then $V=\xi^{(1)}+\mathrm{kern}A^{(0)}+\mathrm{kern}A^{(1)}=\xi^{(1)}+\mathrm{kern}A^{(1)} = S_o$. Hence, in this case, the construction fails.

b) Let $\mathrm{kern}A^{(1)}\subset \mathrm{kern}A^{(0)}$, then $V=\xi^{(1)}+\mathrm{kern}A^{(0)}$ and, for every $x=\xi^{(1)}+u^{(0)}\in V$, a solution y of (220) is given by $y=\xi^{(1)}+h_{11}u^{(0)}$.

c) If $\tilde{\rho} = \mathrm{rank}\tilde{A} = \rho^{(0)}+\rho^{(1)}$, cf. (237) and § 9.4.1, then the projection of any $x \in \mathbb{R}^n$ onto the linear space $\mathrm{kern}A^{(0)} + \mathrm{kern}A^{(1)}$ is equal to x. Hence, $\mathrm{kern}\ A^{(0)}+\mathrm{kern}\ A^{(1)}=\mathbb{R}^n$, and therefore $V = \mathbb{R}^n$, cf. Theorem 9.7.

9.5.2. $\underline{A^{(0)} = 0.}$ The case $A^{(0)}=0$ and $\xi^{(t)}=\xi$, $t=1,\dots,L$. was studied in Theorem 9.6. Thus, we now suppose that the vectors $\xi^{(t)}$, $1\leq t\leq L$, are partially unequal.

For $A^{(0)}=0$, system (246) reads

$$\begin{aligned} &y-h_{tt}x-k^{(t)}=(1-h_{tt})\xi^{(t)}, \quad 1\leq t\leq L, \\ &\qquad k^{(t)}=C^{(t)}p^{(t)}, p^{(t)} \in \mathbb{R}^{n-\rho^{(t)}}, \quad 1\leq t\leq L. \end{aligned} \tag{250}$$

System (250) can be interpreted in different ways.

a) First of all, (250) can be interpreted as a system of $L\cdot n$ linear equations for $(2+L)n - \sum_{t=1}^{L} \rho^{(t)}$ unknowns in $\eta=(x,y,p^{(1)},\dots,p^{(L)})$. For the linear space E of solutions η of the homogeneous system related to (250) we have

$$\dim E \geq 2n - \sum_{t=1}^{L} \rho^{(t)}.$$

Note, that the case $h_{tt} = h_{\tau\tau} \neq 1$ is possible only if

$$\xi^{(t)}-\xi^{(\tau)} \in \mathrm{kern}A^{(t)} + \mathrm{kern}A^{(\tau)}. \tag{251}$$

This, of course, holds if $\xi^{(t)}=\xi^{(\tau)}$ or $\mathrm{kern}A^{(t)} + \mathrm{kern}A^{(\tau)} = \mathbb{R}^n$.

b) For given $h \in H$, $h_{tt}\neq 1$, $1\leq t\leq L$, (250) has a solution y if and only if x is selected such that

$$(h_{tt}-h_{11})x = q^{(1)}-q^{(t)}, \quad t=2,\dots,L \tag{252.1}$$

$$q^{(t)} = k^{(t)}+(1-h_{tt})\xi^{(t)}, \quad t=1,\dots,L. \tag{252.2}$$

Since relation (251) could be observed up to now only in the cases $\xi^{(t)}=\xi^{(\tau)}$ or $\mathrm{kern}A^{(t)} + \mathrm{kern}A^{(\tau)}=\mathbb{R}^n$, cf. sections 9.3.1 and 9.4.1, we suppose now that $h_{tt}\neq h_{\tau\tau}$, $h_{tt}\neq 1$ for all $1\leq t,\tau\leq L, \tau\neq t$. From (252) we then obtain

$$x = \frac{1}{h_{11}-h_{22}} (q^{(2)} - q^{(1)}), \tag{253.1}$$

where $q^{(L)}, q^{(L-1)},\dots,q^{(2)},q^{(1)}$ must fulfill the recursion

$$q^{(t)} = \frac{h_{tt}-h_{t+1t+1}}{h_{t+2t+2}-h_{t+1t+1}} q^{(t+2)} + \frac{h_{t+2t+2}-h_{tt}}{h_{t+2t+2}-h_{t+1t+1}} q^{(t+1)} \tag{253.2}$$

for all $t=L-2,L-1,\dots,2,1$.

We observe that the sum of the coefficients in (253.2) is equal to one for each $t=L-2,\dots,1$. Given $q^{(L-1)},q^{(L)}$, (253.2) yields

$$q^{(t)} = \frac{h_{LL}-h_{tt}}{h_{LL}-h_{L-1L-1}} q^{(L-1)} + \frac{h_{tt}-h_{L-1L-1}}{h_{LL}-h_{L-1L-1}} q^{(L)}, \quad 1\le t\le L-2, \tag{253.3}$$

hence, the vectors $q^{(1)},\dots,q^{(L-2)}$ are affine combinations of $q^{(L-1)}$ and $q^{(L)}$. From (253) we obtain the next theorem:

Theorem 9.10. Let $D=\mathbb{R}^n$, $A^{(0)}=0$ and suppose that (251) does not hold for any pair $1\le t,\tau\le L, t\neq\tau$. The elements x of V are then given by

$$x = \frac{1}{h_{LL}-h_{L-1L-1}} (q^{(L-1)}-q^{(L)}), \tag{254.1}$$

where $h\in H$, $h_{tt}\neq 1$, $h_{tt}\neq h_{\tau\tau}$, $1\le t,\tau\le L$, $\tau\neq t$, and the vectors $k^{(L-1)}\in \mathrm{kern}A^{(L-1)}$, $k^{(L)}\in \mathrm{kern}A^{(L)}$ are selected such that

$$k^{(t)}=q^{(t)}-(1-h_{tt})\xi^{(t)}\in \mathrm{kern}A^{(t)}, \quad t=1,2,\dots,L-2, \tag{254.2}$$

where $q^{(1)},\dots,q^{(L)}$ are defined by (253.3). If $x\in V$, then a solution y of (220) is given by

$$y = \frac{h_{LL}}{h_{LL}-h_{L-1L-1}} q^{(L-1)} - \frac{h_{L-1L-1}}{h_{LL}-h_{L-1L-1}} q^{(L)}. \tag{254.3}$$

Note

a) If $L=2$, then for V we find

$$V = \{\frac{1-h_{11}}{h_{22}-h_{11}} \xi^{(1)} - \frac{1-h_{22}}{h_{22}-h_{11}} \xi^{(2)}: h\in H, h_{11}\neq h_{22}, h_{11}\neq 1, h_{22}\neq 1\}+$$

$$+ \mathrm{kern}A^{(1)} + \mathrm{kern}A^{(2)}.$$

If $x = \frac{1-h_{11}}{h_{22}-h_{11}} \xi^{(1)} - \frac{1-h_{22}}{h_{22}-h_{11}} \xi^{(2)} + u^{(1)} + u^{(2)}$ is an element of V,

then the corresponding solution y of (220) is given by

$$y = h_{22} \frac{1-h_{11}}{h_{22}-h_{11}} \xi^{(1)} - h_{11} \frac{1-h_{22}}{h_{22}-h_{11}} \xi^{(2)} + h_{22}u^{(1)} + h_{11}u^{(2)}.$$

b) Defining $\mu_t = \frac{h_{LL}-h_{tt}}{h_{LL}-h_{L-1L-1}}$, $1 \leq t \leq L-2$, then the conditions (254.2) for $k^{(L-1)} \in \text{kern}A^{(L-1)}$, $k^{(L)} \in \text{kern}A^{(L)}$ read

$$\mu_t A^{(t)}q^{(L-1)} + (1-\mu_t)A^{(t)}q^{(L)} = (1-h_{tt})b^{(t)}, \quad 1 \leq t \leq L-2. \tag{254.2a}$$

9.5.3. <u>$A^{(0)} \neq 0$ and $\xi^{(t)} = \xi$ for all $t=1,\dots,L$.</u> In this case, system (246) reads

$$y = x + k^{(0)}, \quad k^{(0)} \in \text{kern}A^{(0)} \tag{255.1}$$

$$y = \xi + h_{tt}(x-\xi) + k^{(t)}, \quad k^{(t)} \in \text{kern}A^{(t)}, \quad 1 \leq t \leq L. \tag{255.2}$$

Inserting (255.1) into (255.2), for x we find the condition

$$(1-h_{tt})(x-\xi) = k^{(t)}-k^{(0)}, \quad t=1,\dots,L. \tag{256}$$

Supposing that $h_{tt}=\lambda \neq 1$, $t=1,\dots,L$, then

$$x - \xi = \frac{1}{1-\lambda}(k^{(t)}-k^{(0)}), \quad t=1,\dots,L,$$

which means that

$$x \in \xi + \text{kern}A^{(0)} + \bigcap_{t=1}^{L} \text{kern}A^{(t)}. \tag{257}$$

Conversely, if

$$x = \xi + u^{(0)} + v \tag{258.1}$$

fulfills (257), and y is then defined by

$$y = \xi + \lambda u^{(0)} + v, \tag{258.2}$$

then y solves (220). If x,y are related by (258), then $A^{(t)}x \neq A^{(t)}y$ for at least one $1 \leq t \leq L$ if and only if

$$u^{(0)} \notin \bigcap_{t=1}^{L} \text{kern}A^{(t)}.$$

Note, that for L=1 we have again proven Theorem 9.8.

Let $h_{tt} \neq 1$, $1 \leq t \leq L$, and assume now that not all numbers $h_{11},\dots,h_{LL}$ are equal. If $h_{tt} \neq h_{\tau\tau}$ for a pair $t \neq \tau$, then $k^{(\tau)} = k^{(t)}$. Hence,

$$k^{(t)} \in \operatorname{kern} A^{(t)} \cap \operatorname{kern} A^{(\tau)} \tag{259}$$

and the τ-th equation in (256) may be cancelled. Thus, it is sufficient to study the case that $h_{11},\dots,h_{LL}$ are different. Comparing (256) and (252), according to (253) for x we find here the representation

$$x = \xi + \frac{1}{h_{LL}-h_{L-1L-1}} (k^{(L-1)} - k^{(L)}), \tag{260.1}$$

where $k^{(t)} \in \operatorname{kern} A^{(t)}$, $t=L-1,L$, must fulfill the relations

$$k^{(t)} = \frac{h_{LL}-h_{tt}}{h_{LL}-h_{L-1L-1}} k^{(L-1)} + \frac{h_{tt}-h_{L-1L-1}}{h_{LL}-h_{L-1L-1}} k^{(L)} \in \operatorname{kern} A^{(t)} \tag{260.2}$$

for all $t=0,1,\dots,L-2$, where $h_{oo}=1$.

Moreover, if x is defined by (260.1), (260.2), then

$$y = \xi + \frac{h_{LL}}{h_{LL}-h_{L-1L-1}} k^{(L-1)} - \frac{h_{L-1L-1}}{h_{LL}-h_{L-1L-1}} k^{(L)} \tag{260.3}$$

is a solution of (220).

Remark

a) The above considerations yield that the n-vectors $k^{(L-1)}$, $k^{(L)}$ have to fulfill the following homogeneous system of $\sum_{t=0}^{L} \rho^{(t)}$ linear equations

$$(h_{LL}-h_{tt})A^{(t)}k^{(L-1)} + (h_{tt}-h_{L-1L-1})A^{(t)}k^{(L)} = 0, \quad t=0,1,\dots,L,$$

cf. (254.2a).

b) In the special case L=2, system (260) reads

$$x = \xi + u^{(1)} + u^{(2)}$$

$$y = \xi + h_{22}u^{(1)} + h_{11}u^{(2)},$$

where the n-vectors $u^{(1)}$, $u^{(2)}$ must be chosen such that

$$(1-h_{22})A^{(0)}u^{(1)} + (1-h_{11})A^{(0)}u^{(2)} = 0,$$
$$A^{(1)}u^{(1)} = 0,\ A^{(2)}u^{(2)} = 0.$$

In order to describe now V, let $H_1=\{h\in H:\ h_{11}= \dots = h_{LL} \neq 1\}$, $H_2=\{h\in H$: all $h_{11},\dots,h_{LL}$ are different from each other and different from one$\}$, $H_3=\{h\in H: h_{tt}\neq 1,\ 1\le t\le L,\ h\notin H_1\cup H_2\}$. According to (257), (258.1), for $h\in H_1$ the set of solutions x of (256) is given by

$$V_1 = \xi + \mathrm{kern}A^{(0)} + \bigcap_{t=1}^{L} \mathrm{kern}A^{(t)}.$$

If $h\in H_2$, the set V_2 of solutions x of (256) is then represented by (260.1), (260.2). Finally, for $h\in H_3$ the set V_3 of solutions x of (256) may be described by combining (259) and (260.1), (260.2).

Thus, we finally obtain this result:

Theorem 9.11. Let $D=\mathbb{R}^n$, $A^{(0)}\neq 0$ and $\xi^{(t)}=\xi$ for all $t=1,\dots,L$, then $V=V_1\cup V_2\cup V_3$.

9.6. Solutions of (192.4)-(192.6) in case $D \neq \mathbb{R}^n$. Let S_D denote the set of D-stationary points, cf. Definition 4.1 and Theorem 7.4. If S_o is again defined by (221.2), then, using Lemma 9.1, we have the inclusion

$$D\cap S_o \subset S_D.$$

Furthermore, if, for every $x\in D$, (195) has a solution $y\in D$, then $S_D=D\cap S_o$.

9.6.1. The case $x\in \mathring{D}$; Inclusions for S_D. If x is an interior of D, which lies also in $V\setminus S_o$, where V is defined by (223), then, according to section 9.3, there is a tuple (y,B,h) solving (192.1)-(192.5) such that $A^{(t)}y\neq A^{(t)}x$ for at least one $1\le t\le L$. Since $d=y-x$ is also a feasible direction for D at x, x is not D-stationary. Denoting by ∂D boundary of D, we therefore find

$$S_D \subset D \setminus (\overset{o}{D} \cap (V \setminus S_o)) = \partial D \cup (D \setminus V) \cup (D \cap S_o).$$

Thus, we have the following result:

Theorem 9.12. a) If $x \in \overset{o}{D} \cap (V \setminus S_o)$, then (192.1)-(192.5) has a solution (y,B,h) such that $A^{(t)}y \neq A^{(t)}x$ for at least one $1 \leq t \leq L$, and $d=y-x$ is a feasible direction for D at x. b) For every feasible set D it is

$$S_D \subset \partial D \cup (D \setminus V) \cup (D \cap S_o). \tag{261}$$

Remark

If $V = \mathbb{R}^n$, cf. § 9.3 - 9.5, then

$$S_D \subset \partial D \cup (D \cap S_o).$$

If D is open, $D \setminus V = \emptyset$ and $D \cap S_o = \emptyset$, then (261) yields $S_o = \emptyset$. Hence, according to Lemma 4.1 we have this consequence:

Corollary 9.2. Let D be open and suppose that $D \setminus V = \emptyset$, $D \cap S_o = \emptyset$, then the basic stochastic optimization problem (SOP) min F(x) s.t. $x \in D$ has no optimal solution for each loss function u such that F is constant on a line segment xy of $\mathbb{R}^n$ if and only if $A^{(t)}y = A^{(t)}x$ for all $1 \leq t \leq L$.

9.6.2. Further inclusions for S_D. We suppose here that the system of linear equations (195), i.e.

$$A^{(0)}y = A^{(0)}x,\ A^{(t)}y = b^{(t)},\ t=1,\dots,L \tag{262}$$

has a solution $y \in \mathbb{R}^n$ for every $x \in D$.

In order to derive then sharper inclusions for S_D, we introduce the notion of an (A,b)-boundary point of D.

Definition 9.1. Suppose that (262) holds. A point $x \in D \setminus S_o$ is called an (A,b)-boundary point of D if there is no solution y of (195) such that $d=y-x$ is a feasible direction for D at x. Let $\partial_{(A,b)}D$ denote the set of (A,b)-boundary points of D.

Justifying this notion, let $x \in \mathring{D} \setminus S_0$ and consider any (existing) solution $y \in \mathbb{R}^n$ of (195), cf. (262). Since x is an interior point of D and $A^{(t)}x \neq b^{(t)}$ for at least one $1 \leq t \leq L$, we have $y \neq x$ and $d = y - x$ is a feasible direction for D at x. Hence, by Definition 9.1, x is not an element of $\partial_{(A,b)}D$. Consequently

$$\mathring{D} \setminus S_0 \subset D \setminus \partial_{(A,b)}D$$

which yields

$$\partial_{(A,b)}D \subset \partial D \cup (D \cap S_0).$$

Since $\partial_{(A,b)}D \cap S_0 = \emptyset$, see Definition 9.1, we find this lemma:

Lemma 9.4. Assume that (262) holds true, then

$$\partial_{(A,b)}D \subset \partial D. \tag{263}$$

Consider now a point $x \in D \setminus S_0$. If $x \notin \partial_{(A,b)}D$, then there is a solution y of (195) such that $d = y - x$ is a feasible direction for D at x. Furthermore, because of $x \notin S_0$ and $y \in S_0$, we have $A^{(t)}y \neq A^{(t)}x$ for at least one $1 \leq t \leq L$. Defining $B = \alpha 1_r'$, $h = 0$, we find that the tuple (y, B, h) solves (192.1)-(192.5), hence it is $x \notin S_D$.

Supplementary to (261) we obtain

Theorem 9.13. Assume that (162) holds true, then

$$S_D \subset \partial_{(A,b)}D \cup (D \cap S_0). \tag{264}$$

Proof. The above considerations yield

$$(D \setminus S_0) \setminus \partial_{(A,b)}D \subset D \setminus S_D,$$

and therefore,

$$D \setminus ((D \setminus S_0) \setminus \partial_{(A,b)}D) \supset D \setminus (D \setminus S_D) = S_D,$$

hence $S_D \subset (D \cap S_0) \cup \partial_{(A,b)}D$.

Note. If (262) holds, then (putting h=0) we find that $D \subset V$, cf. (223). Hence, (261) yields $S_D \subset \partial D \cup (D \cap S_0)$.

It is easy to characterize the domains D having property (262). Indeed, according to (262), for every x we have the representation

$x = y + k^{(0)}$, where $y \in S_o, k^{(0)} \in \text{kern}A^{(0)}$.

Hence, we obtain the following lemma:

Lemma 9.5. Property (262) holds true if and only if $D \subset S_o + \text{kern}A^{(0)}$.

9.6.3. Solutions of (192.4)-(192.6) for special feasible sets D. Let $h=(h_{11},\dots,h_{LL})'$ be an arbitrary given element of $H \setminus \{1\}$.

In this section we consider the conditions (192.4)-(192.6) for some special selections of the feasible domain D.

9.6.3.1. Affine subspaces. Assume that D is defined by

$$D = \{x \in \mathbb{R}^n : Gx=g\}, \tag{265}$$

where G is a p×n matrix and g is a p-vector. Let $\gamma=\text{rank}G=\text{rank}(G,g)<n$ and select a $\gamma \times n$ submatrix $\tilde{G}$ of G having $\text{rank}\tilde{G}=\gamma$. If (265) holds, then for every $x \in D$, system (192.4)-(192.6) has the form

$$\hat{A}^{(0)}y = \hat{A}^{(0)}x \tag{266.1}$$

$$A^{(t)}y = b^{(t)} + h_{tt}(A^{(t)}x-b^{(t)}, \quad t=1,\dots,L, \tag{266.2}$$

where the $(\rho^{(0)}+\gamma)\times n$ matrix $\hat{A}^{(0)}$ is defined by

$$\hat{A}^{(0)} = \begin{pmatrix} \tilde{A}^{(0)} \\ G \end{pmatrix} .$$

Thus, (266) has exactly the same form as system (220) for the case $D=\mathbb{R}^n$, cf. § 9.3. The only difference is that $\hat{A}^{(0)}$ has more rows than $\tilde{A}^{(0)}$. This means, of course, that all results obtained in sections 9.3-9.5 for $D=\mathbb{R}^n$ can be easily transfered to the present case (265) by simply replacing $\tilde{A}^{(0)}(A^{(0)}$, resp.) by $\hat{A}^{(0)}$. To give one example, corresponding to Theorem 9.4, we have this theorem:

Theorem 9.14. Let D be given by (265), and suppose that

$$\operatorname{rank}\begin{pmatrix} \tilde{A}^{(0)} \\ \tilde{G} \\ A^{(1)} \\ \cdot \\ \cdot \\ \cdot \\ A^{(L)} \end{pmatrix} = \rho^{(0)} + \gamma + Lm, \tag{267}$$

then (192.4)-(192.6) has a solution $y(\in D)$ for every $x \in D$ and every $h \in H$. Moreover,

$S_D = D \cap S_o = \{x \in \mathbb{R}^n : Gx=g,\ A^{(t)}x=b^{(t)},\ t=1,\dots,L\}$

is an $(n-\gamma-Lm)$-dimensional linear submanifold of D.

9.6.3.2. Convex polyhedrons. Assume that D is given by

$$D = \{x \in \mathbb{R}^n : Gx=g,\ Kx \le k\}, \tag{268}$$

where (G,g) is defined as in (265) and K is a $q \times n$ matrix, $k \in \mathbb{R}^q$. This case (268) can be handled by the following two methods:

a) Introducing slack-variables $\delta \in \mathbb{R}^q$ into the inequality $Kx \le k$, for every $x \in D$ conditions (192.4)-(192.6) read

$$\tilde{A}^{(0)}y = \tilde{A}^{(0)}x \tag{269.1}$$

$$\tilde{G}y = \tilde{G}x \tag{269.2}$$

$$Ky = k-\delta,\ \delta \ge 0 \tag{269.3}$$

$$A^{(t)}y = b^{(t)} + h_{tt}(A^{(t)}x-b^{(t)}),\ t=1,\dots,L, \tag{269.4}$$

where $\tilde{A}^{(0)}$, $\tilde{G}$ are defined as before.

Interpreting now $\delta \ge 0$ as an arbitrary, but fixed q-vector of nonnegative parameters, system (269) has again the same form as (220). Hence, the results of sections 9.3-9.5 may be transfered to this case by simply replacing $\tilde{A}^{(0)}$ ($A^{(0)}$, resp.) by the matrix

$$\begin{pmatrix} \tilde{A}^{(0)} \\ \tilde{G} \\ K \end{pmatrix},$$

cf. 9.6.3.1.

b) The second method combines the construction procedures described in sections 9.3-9.5 (for $D=\mathbb{R}^n$) and section 9.6.1(for $x \in \overset{o}{D}$). Let x be a given element of D, hence $Gx=g$, $Kx\leq k$, and define the following index sets

$$I_1(x) = \{i: 1\leq i\leq q, K_i x=k_i\},$$
$$I_2(x) = \{i: 1\leq i\leq q: K_i x<k_i\}. \tag{270}$$

Solving (192.4)-(192.6), in the first step we consider the reduced system

$$\hat{A}^{(0)}y = \hat{A}^{(0)}x \tag{271.1}$$

$$A^{(t)}y = b^{(t)} + h_{tt}(A^{(t)}x-b^{(t)}), \ t=1,\dots,L, \tag{271.2}$$

where, in the present case, $\hat{A}^{(0)}$ is defined by

$$\hat{A}^{(0)} = \begin{pmatrix} \tilde{A}^{(0)} \\ \tilde{G} \\ K_I \end{pmatrix}, \tag{272}$$

$K_I = (K_i)_{i \in I_1(x)}$, and K_i denotes the i-th row of K, cf. (266) and (269).

Suppose now that by transfering one of the construction methods from sections 9.3-9.5 to (271), we have already determined a solution $y\neq x$ of (271). Consider then the vector

$$y(\mu) = x + \mu(y-x), \ \mu \in \mathbb{R}. \tag{273}$$

Because of (271.1) and (272) we have

$$\tilde{A}^{(0)}y(\mu) = \tilde{A}^{(0)}x \tag{274.1}$$

$$\tilde{G}y(\mu) = \tilde{A}^{(0)}x \tag{274.2}$$

$$K_I y(\mu) = K_I x \tag{274.3}$$

for all $\mu \in \mathbb{R}$. Moreover, utilizing (271.2), we find for all $\mu \in \mathbb{R}$

$$A^{(t)}y(\mu) = A^{(t)}(x+\mu(y-x)) = \tag{274.4}$$

$$= A^{(t)}x+\mu(b^{(t)}+h_{tt}(A^{(t)}x-b^{(t)}) - A^{(t)}x) =$$

$$= b^{(t)} + (\mu h_{tt}+(1-\mu))(A^{(t)}x-b^{(t)}).$$

According to (274) we see that for all $\mu \in \mathbb{R}$ the tuple

$$(y(\mu),h(\mu)), \text{ where } h(\mu) = \mu h+(1-\mu)1_L, \tag{275}$$

fulfills relations (192.4)-(192.5) as well as (274.2) and (274.3) for all $\mu \in \mathbb{R}$. Since for the given $h \in H \setminus \{1_L\}$ there is a matrix B such that (B,h) fulfills (192.1)-(192.3), and - of course - also $(I,1_L)$ solves the linear relations (192.1)-(192.3), the tuple

$$(B(\mu),h(\mu)), \text{ where } B(\mu) = \mu B + (1-\mu)I, \tag{276}$$

solves (192.1)-(192.3) for every $0 \leq \mu \leq 1$. Consequently, the tuple

$$(y(\mu),B(\mu),h(\mu))$$

satisfies (192.1)-(192.5) as well as (274.2) and (274.3) for all $0 \leq \mu \leq 1$. Since $h \neq 1_L$, we have

$$h(\mu) \neq 1_L \text{ for every } 0<\mu \leq 1.$$

Furthermore, if $h_{tt} \neq 1$ for every $1 \leq t \leq L$, then

$$h_{tt}(\mu) = \mu h_{tt} + (1+\mu) \neq 1 \text{ for all } 1 \leq t \leq L \text{ and } 0<\mu \leq 1.$$

Finally, we have to satisfy the remaining conditions

$$K_i y(\mu) \leq k_i \text{ for all } i \in I_2(x),$$

cf. (270). Because of $K_i x<k_i$ for all $i \in I_2(x)$ and

$$K_i y(\mu) = K_i x + \mu(K_i y-K_i x)$$

we may select a number $0<\mu_o \leq 1$ such that

$$K_i y(\mu) \leq k_i \text{ for all } i \in I_2(x) \text{ and each } 0 \leq \mu \leq \mu_o.$$

Summarizing the above considerations, we reach the final theorem:

Theorem 9.15. Suppose that D is given by (268), and let x be an element of D. For a given vector $h \in H \setminus \{1_L\}$ determine a solution y of (271) and define the tuple $(y(\mu), B(h), h(\mu))$ by (273), (275) and (276), where B is the matrix related to h. Then, there is a number $0<\mu_0\leq 1$ such that $(y(\mu), B(\mu), h(\mu))$ satisfies (192.1)-(192.6) for every $0\leq\mu\leq\mu_0$. If $A^{(t)}x \neq b^{(t)}$ for an index $t=t_0$, $1\leq t_0\leq L$, and $h_{t_0 t_0} \neq 1$, then $A^{(t_0)}y(\mu) \neq A^{(t_0)}x$ for all $0<\mu\leq\mu_0$.

REFERENCES

[1] Abel, P.: Stochastische Optimierung bei partieller Information, Mathematical Systems in Economics, Vol. 96, Königstein/Ts., A. Hain 1984

[2] Aitchison, J.; Dunsmore, I.R.: Statistical Prediction Analysis, Cambridge, University Press 1975

[3] Berman, A.; Plemmons, R.J.: Nonnegative matrices in the mathematical sciences, New York-London, Academic Press 1979

[4] Birge, J.R.; Wets, R.J.-B.: Designing Approximation Schemes for Stochastic Optimization Problems, in particular for Stochastic Programs with Recourse, Math. Programming Study 27, 54-102 (1986)

[5] Birge, J.R.; Wets, R.J.-B.: A sublinear approximation method for stochastic programming, University of Michigan, Dept. of Industrial and Operations Engineering, Technical Report 86-26, July 1986

[6] Blackwell, D.: Equivalent Comparisons of Experiments, Ann.Math. Statist. 24, 265-272 (1953)

[7] Blum, E.; Oettli, W.: Mathematische Optimierung, Econometrics and Operations Research, Vol. XX, Berlin-Heidelberg-New York, Springer-Verlag 1975

[8] Blum, J.R.; Rosenblatt, J.: On partial a priori information in statistical inference, Ann. Math. Statist. 38, 1671-1678 (1967)

[9] Cartier, P.; Fell, J.M.G.; Meyer, P.-A.: Comparison des mesures portées par un ensemble convexe compact, Bull.Soc.Math. France 92, 435-445 (1964)

[10] De Groot, M.H.: Uncertainty, information and sequential experiments, Ann.Math. Statist. 33, 404-419 (1962)

[11] De Groot, M.H.: Changes in utility as information, Theory and Decisions 17, 287-303 (1984)

[12] Dunford, N.; Schwartz, J.T.: Linear Operators, Part I, New York, J. Wiley 1957

[13] Dupacova, J.: Minimax stochastic programs with nonconvex nonseparable penalty functions, Colloquia Mathematica Societatis Janos Bolyai, 12. Progress in Operations Research, Eger (Hungary) 1974, 303-316

[14] Ermoliev, Y.; Gaivoronski, A.: Stochastic Quasigradient Methods and their Implementation. IIASA Working Paper, Laxenburg 1983

[15] Ermoliev, Y.: Stochastic Quasigradient Methods and their Application to System Optimization, Stochastics 9, 1-36 (1983)

[16] Ermoliev, Y.; Gaivoronski, A.; Nedeva, C.: Stochastic Optimization Problems with Incomplete Informations on Distribution Functions, SIAM J. Control and Optimization 23, 697-716 (1985)

[17] Ferguson, Th. S.: Mathematical Statistics, New York-London, Academic Press 1967

[18] Fishburn, P.C.: Stochastic Dominance and Moments of Distributions, Math. Operations Research 5, 94-100 (1980)

[19] Frauendorfer, K.; Kall, P.: A solution method for SLP recourse problems with arbitrary multivariate distributions - The independent case, Universität Zürich, Institut für Operations Research, Manuskript, Juli 1986

[20] Fulkerson, D.R.: Hitchcock transportation problem, Rand Corp. Rep. P 890, Santa Monica 1956

[21] Hardy, G.H.; Littlewood, J.E.; Polya, G.: Inequalities, Cambridge, University Press 1973

[22] Hinderer, K.: Foundations of Non-stationary Dynamic Programming with Discrete Time Parameter, Lecture Notes in Operations Research and Mathematical Systems, Vol. 33, Berlin-Heidelberg-New York, Springer-Verlag 1970

[23] Hiriart-Urruty, J.B.: Contributions a la programmation mathematique: Cas deterministe et stochastique, Thèse, Université de Clermont-Ferrand II, 1977

[24] Kall, P.: Qualitative Aussagen zu einigen Problemen der stochastischen Programmierung, Z. Wahrscheinlichkeitstheorie verw. Gebiete 6, 246-272 (1966)

[25] Kall, P.: Approximations to Stochastic Programs with Complete Fixed Recourse, Numerische Mathematik 22, 333-339 (1974)

[26] Kall, P.: Stochastic Linear Programming, Econometrics and Operations Research, Vol XXI, Berlin-Heidelberg-New York, Springer-Verlag 1976

[27] Kall, P.: Computational Methods for Solving Two-stage Stochastic Linear Programming Problems, Journal of Applied Mathematics and Physics (ZAMP) 30, 261-271 (1979)

[28] Kall, P.: Stochastic Programming, European J. of Operations Research 10, 125-130 (1982)

[29] Kall, P.: Approximations to optimization problems: An elementary review, Mathematics of Operations Research 11, 9-18 (1986)

[30] Kall, P.: Stochastic programs with recourse: An upper bound and the related moment problem, Universität Zürich, Institut für Operations Research, Manuskript, August 1986

[31] Kesten, H.: Accelerated stochastic approximation, Ann.Math. Statist. 29, 41-59 (1958)

[32] Kullback, S.: Information Theory and Statistics, New York-London, J. Wiley 1959

[33] Kushner, H.J.; Clark, D.S.: Stochastic Approximation Methods for Constrained and Unconstrained Systems, New York-Heidelberg-Berlin, Springer-Verlag 1978

[34] Lindley, D.V.: Making Decisions, London-New York-Brisbane-Toronto -Singapore, J. Wiley 1985

[35] Luenberger, D.G.: Optimization by vector space methods, New York-London-Sydney-Toronto, J. Wiley 1969

[36] Markowitz, H.M.: Portfolio selections, New York-London, J. Wiley 1959

[37] Marti, K.: Approximationen der Entscheidungsprobleme mit linearer Ergebnisfunktion und positiv homogener, subadditiver Verlustfunktion, Z. Wahrscheinlichkeitstheorie verw. Gebiete 31, 203-233 (1975)

[38] Marti, K.: Stochastische Dominanz und stochastische lineare Programme, Methods of Operations Research 23, 141-160 (1977)

[39] Marti, K.: Diskretisierung stochastischer Programme unter Berücksichtigung der Problemstruktur, ZAMM 50, T105-T108 (1979)

[40] Marti, K.: Approximationen stochastischer Optimierungsprobleme, Mathematical Systems in Economics, Vol. 43, Königstein/Ts., A. Hain 1979

[41] Marti, K.: Solving stochastic linear programs by semi-stochastic approximation algorithms. In: Kall, P.; Prekopa, A. (eds.): Recent results in stochastic programming, Lecture Notes in Economics and Mathematical Systems, Vol. 179 (1980), pp. 191-213

[42] Marti, K.: On stochastic dominance and the construction of directions of decrease in stochastic programs having a discrete distribution, Methods of Operations Research 41, 175-178 (1981)

[43] Marti, K.: Computation of Descent Directions in Stochastic Optimization Problems with Invariant Distributions. ZAMM 65, 355-378 (1985)

[44] Marti, K.; Fuchs, E.: Computation of Descent Directions and Efficient Points in Stochastic Optimization Problems Without Using Derivatives, Math. Programming Study 28, 132-156 (1986)

[45] Marti, K.; Fuchs, E.: Rates of Convergence of Semi-Stochastic Approximation Procedures for Solving Stochastic Optimization Problems, Optimization 17, 243-265 (1986)

[46] Nazareth, J.L.; Wets, R.J.-B.: Algorithms for Stochastic Programs: The case of nonstochastic tenders, Math. Programming Study 28, 1-28 (1986)

[47] Pflug, G. CH.: On the determination of the step size in stochastic quasigradient methods, IIASA Laxenburg, Collaborative Paper CP-83-25, 1983

[48] Poljak, B.T.; Tsypkin, Ya.Z.: Robust pseudogradient adaptation algorithms, Automation and Remote Control 41, 1404-1410 (1980)

[49] Poljak, B.T.; Tsypkin, Ya.Z.: Optimal pseudogradient adaptation algorithms, Automation and Remote Control 41, 1101-1110 (1980)

[50] Raiffa, H.; Schlaifer, R.: Applied Statistical Decision Theory, Studies in Managerial Economics, Division of Research, Graduate School of Business Administration, Harvard University, Boston, 1961

[51] Sherman, S.: On a theorem of Hardy, Littlewood, Polya, and Blackwell, Proc.Nat.Acad. Sci. USA 37, 826-831 (1951)

[52] Smith, C.R.; Grandy, W.T. (eds.): Maximum-Entropy and Bayesian Mathods in Inverse Problems, Dordrecht-Boston-Lancaster, D. Reidel 1985

[53] Sorenson, H.W.: Parameter estimation, New York-Basel, Marcel Dekker 1980

[54] Stoer, J.; Witzgall, Ch.: Convexity and Optimization in Finite Dimensions I, Grundlehren der mathematischen Wissenschaften in Einzeldarstellungen, Band 163, Berlin-Heidelberg-New York, Springer-Verlag 1970

[55] Stoyan, D.: Qualitative Eigenschaften und Abschätzungen stochastischer Modelle, München-Wien, R. Oldenbourg 1977

[56] Strazicky, B.: On the Algorithm for Solution of the Two-Stage Stochastic Programming Problem. Methods of Operations Research 19, 142-156 (1974)

[57] Theil, H.: Economic Forecasts and Policy, Contributions to Economics Analysis XV, Amsterdam-Oxford, North Holland Publ.Co. 1975

[58] Tribus, M.: Planungs- und Entscheidungstheorie ingenieurwissenschaftlicher Probleme, Braunschweig, Vieweg 1973

[59] Tsypkin, Ya.Z.: Foundations of the Theory of Learning Systems, New York-London, Academic Press 1973

[60] Van Slyke, R.M.; Wets, R.: L-Shaped Linear Programs With Applications To Optimal Control and Stochastic Programming, SIAM J. Appl.Math, 17, 638-663 (1969)

[61] Vickson, R.G.; Ziemba, W.T. (eds.): Stochastic Optimization Models in Finance, New York-London, Academic Press 1975

[62] Wasan, M.T.: Stochastic Approximation. Cambridge, University Press 1969

[63] Wets, R.: Stochastic Programming: Solutions Techniques and Approximation Schemes. In: Bachem, A.; Grötschel, M.; Korte, B.: Mathematical Programming - The State of the Art, Berlin-Heidelberg-New York-Tokyo, Springer-Verlag 1983, pp. 566-603

[64] Wets, R.: Stochastic programs with fixed recourse: the equivalent deterministic program, SIAM Review 16, 309-339 (1974)

[65] Wets, R.J.-B.: Solving Stochastic Programs with Simple Recourse, Stochastics 10, 219-242 (1983)

[66] Wets, R.J.-B.: Large Scale Linear Programming Techniques in Stochastic Programming. IIASA WP-84-90, November 1984

[67] Bibliography on stochastic optimization, to be published in "Numerical Techniques for Stochastic Optimization Problems", Springer Verlag

INDEX

Vol. 211: P. van den Heuvel, The Stability of a Macroeconomic System with Quantity Constraints. VII, 169 pages. 1983.

Vol. 212: R. Sato and T. Nôno, Invariance Principles and the Structure of Technology. V, 94 pages. 1983.

Vol. 213: Aspiration Levels in Bargaining and Economic Decision Making. Proceedings, 1982. Edited by R. Tietz. VIII, 406 pages. 1983.

Vol. 214: M. Faber, H. Niemes und G. Stephan, Entropie, Umweltschutz und Rohstoffverbrauch. IX, 181 Seiten. 1983.

Vol. 215: Semi-Infinite Programming and Applications. Proceedings, 1981. Edited by A.V. Fiacco and K.O. Kortanek. XI, 322 pages. 1983.

Vol. 216: H.H. Müller, Fiscal Policies in a General Equilibrium Model with Persistent Unemployment. VI, 92 pages. 1983.

Vol. 217: Ch. Grootaert, The Relation Between Final Demand and Income Distribution. XIV, 105 pages. 1983.

Vol. 218: P. van Loon, A Dynamic Theory of the Firm: Production, Finance and Investment. VII, 191 pages. 1983.

Vol. 219: E. van Damme, Refinements of the Nash Equilibrium Concept. VI, 151 pages. 1983.

Vol. 220: M. Aoki, Notes on Economic Time Series Analysis: System Theoretic Perspectives. IX, 249 pages. 1983.

Vol. 221: S. Nakamura, An Inter-Industry Translog Model of Prices and Technical Change for the West German Economy. XIV, 290 pages. 1984.

Vol. 222: P. Meier, Energy Systems Analysis for Developing Countries. VI, 344 pages. 1984.

Vol. 223: W. Trockel, Market Demand. VIII, 205 pages. 1984.

Vol. 224: M. Kiy, Ein disaggregiertes Prognosesystem für die Bundesrepublik Deutschland. XVIII, 276 Seiten. 1984.

Vol. 225: T.R. von Ungern-Sternberg, Zur Analyse von Märkten mit unvollständiger Nachfragerinformation. IX, 125 Seiten. 1984

Vol. 226: Selected Topics in Operations Research and Mathematical Economics. Proceedings, 1983. Edited by G. Hammer and D. Pallaschke. IX, 478 pages. 1984.

Vol. 227: Risk and Capital. Proceedings, 1983. Edited by G. Bamberg and K. Spremann. VII, 306 pages. 1984.

Vol. 228: Nonlinear Models of Fluctuating Growth. Proceedings, 1983. Edited by R.M. Goodwin, M. Krüger and A. Vercelli. XVII, 277 pages. 1984.

Vol. 229: Interactive Decision Analysis. Proceedings, 1983. Edited by M. Grauer and A.P. Wierzbicki. VIII, 269 pages. 1984.

Vol. 230: Macro-Economic Planning with Conflicting Goals. Proceedings, 1982. Edited by M. Despontin, P. Nijkamp and J. Spronk. VI, 297 pages. 1984.

Vol. 231: G.F. Newell, The M/M/∞ Service System with Ranked Servers in Heavy Traffic. XI, 126 pages. 1984.

Vol. 232: L. Bauwens, Bayesian Full Information Analysis of Simultaneous Equation Models Using Integration by Monte Carlo. VI, 114 pages. 1984.

Vol. 233: G. Wagenhals, The World Copper Market. XI, 190 pages. 1984.

Vol. 234: B.C. Eaves, A Course in Triangulations for Solving Equations with Deformations. III, 302 pages. 1984.

Vol. 235: Stochastic Models in Reliability Theory. Proceedings, 1984. Edited by S. Osaki and Y. Hatoyama. VII, 212 pages. 1984.

Vol. 236: G. Gandolfo, P.C. Padoan, A Disequilibrium Model of Real and Financial Accumulation in an Open Economy. VI, 172 pages. 1984.

Vol. 237: Misspecification Analysis. Proceedings, 1983. Edited by T.K. Dijkstra. V, 129 pages. 1984.

Vol. 238: W. Domschke, A. Drexl, Location and Layout Planning. IV, 134 pages. 1985.

Vol. 239: Microeconomic Models of Housing Markets. Edited by K. Stahl. VII, 197 pages. 1985.

Vol. 240: Contributions to Operations Research. Proceedings, 1984. Edited by K. Neumann and D. Pallaschke. V, 190 pages. 1985.

Vol. 241: U. Wittmann, Das Konzept rationaler Preiserwartungen. XI, 310 Seiten. 1985.

Vol. 242: Decision Making with Multiple Objectives. Proceedings, 1984. Edited by Y.Y. Haimes and V. Chankong. XI, 571 pages. 1985.

Vol. 243: Integer Programming and Related Areas. A Classified Bibliography 1981–1984. Edited by R. von Randow. XX, 386 pages. 1985.

Vol. 244: Advances in Equilibrium Theory. Proceedings, 1984. Edited by C.D. Aliprantis, O. Burkinshaw and N.J. Rothman. II, 235 pages. 1985.

Vol. 245: J.E.M. Wilhelm, Arbitrage Theory. VII, 114 pages. 1985.

Vol. 246: P.W. Otter, Dynamic Feature Space Modelling, Filtering and Self-Tuning Control of Stochastic Systems. XIV, 177 pages. 1985.

Vol. 247: Optimization and Discrete Choice in Urban Systems. Proceedings, 1983. Edited by B.G. Hutchinson, P. Nijkamp and M. Batty. VI, 371 pages. 1985.

Vol. 248: Plural Rationality and Interactive Decision Processes. Proceedings, 1984. Edited by M. Grauer, M. Thompson and A.P. Wierzbicki. VI, 354 pages. 1985.

Vol. 249: Spatial Price Equilibrium: Advances in Theory, Computation and Application. Proceedings, 1984. Edited by P. T. Harker. VII, 277 pages. 1985.

Vol. 250: M. Roubens, Ph. Vincke, Preference Modelling. VIII, 94 pages. 1985.

Vol. 251: Input-Output Modeling. Proceedings, 1984. Edited by A. Smyshlyaev. VI, 261 pages. 1985.

Vol. 252: A. Birolini, On the Use of Stochastic Processes in Modeling Reliability Problems. VI, 105 pages. 1985.

Vol. 253: C. Withagen, Economic Theory and International Trade in Natural Exhaustible Resources. VI, 172 pages. 1985.

Vol. 254: S. Müller, Arbitrage Pricing of Contingent Claims. VIII, 151 pages. 1985.

Vol. 255: Nondifferentiable Optimization: Motivations and Applications. Proceedings, 1984. Edited by V.F. Demyanov and D. Pallaschke. VI, 350 pages. 1985.

Vol. 256: Convexity and Duality in Optimization. Proceedings, 1984. Edited by J. Ponstein. V, 142 pages. 1985.

Vol. 257: Dynamics of Macrosystems. Proceedings, 1984. Edited by J.-P. Aubin, D. Saari and K. Sigmund. VI, 280 pages. 1985.

Vol. 258: H. Funke, Eine allgemeine Theorie der Polypol- und Oligopolpreisbildung. III, 237 pages. 1985.

Vol. 259: Infinite Programming. Proceedings, 1984. Edited by E.J. Anderson and A.B. Philpott. XIV, 244 pages. 1985.

Vol. 260: H.-J. Kruse, Degeneracy Graphs and the Neighbourhood Problem. VIII, 128 pages. 1986.

Vol. 261: Th.R. Gulledge, Jr., N.K. Womer, The Economics of Made-to-Order Production. VI, 134 pages. 1986.

Vol. 262: H.U. Buhl, A Neo-Classical Theory of Distribution and Wealth. V, 146 pages. 1986.

Vol. 263: M. Schäfer, Resource Extraction and Market Structure. XI, 154 pages. 1986.

Vol. 264: Models of Economic Dynamics. Proceedings, 1983. Edited by H.F. Sonnenschein. VII, 212 pages. 1986.

Vol. 265: Dynamic Games and Applications in Economics. Edited by T. Başar. IX, 288 pages. 1986.

Vol. 266: Multi-Stage Production Planning and Inventory Control. Edited by S. Axsäter, Ch. Schneeweiss and E. Silver. V, 264 pages. 1986.

Vol. 267: R. Bemelmans, The Capacity Aspect of Inventories. IX, 165 pages. 1986.

Vol. 268: V. Firchau, Information Evaluation in Capital Markets. VII, 103 pages. 1986.

Vol. 269: A. Borglin, H. Keiding, Optimality in Infinite Horizon Economies. VI, 180 pages. 1986.

Vol. 270: Technological Change, Employment and Spatial Dynamics. Proceedings 1985. Edited by P. Nijkamp. VII, 466 pages. 1986.

Vol. 271: C. Hildreth, The Cowles Commission in Chicago, 1939–1955. V, 176 pages. 1986.

Vol. 272: G. Clemenz, Credit Markets with Asymmetric Information. VIII, 212 pages. 1986.

Vol. 273: Large-Scale Modelling and Interactive Decision Analysis. Proceedings, 1985. Edited by G. Fandel, M. Grauer, A. Kurzhanski and A.P. Wierzbicki. VII, 363 pages. 1986.

Vol. 274: W.K. Klein Haneveld, Duality in Stochastic Linear and Dynamic Programming. VII, 295 pages. 1986.

Vol. 275: Competition, Instability, and Nonlinear Cycles. Proceedings, 1985. Edited by W. Semmler. XII, 340 pages. 1986.

Vol. 276: M.R. Baye, D.A. Black, Consumer Behavior, Cost of Living Measures, and the Income Tax. VII, 119 pages. 1986.

Vol. 277: Studies in Austrian Capital Theory, Investment and Time. Edited by M. Faber. VI, 317 pages. 1986.

Vol. 278: W.E. Diewert, The Measurement of the Economic Benefits of Infrastructure Services. V, 202 pages. 1986.

Vol. 279: H.-J. Büttler, G. Frei and B. Schips, Estimation of Disequilibrium Models. VI, 114 pages. 1986.

Vol. 280: H.T. Lau, Combinatorial Heuristic Algorithms with FORTRAN. VII, 126 pages. 1986.

Vol. 281: Ch.-L. Hwang, M.-J. Lin, Group Decision Making under Multiple Criteria. XI, 400 pages. 1987.

Vol. 282: K. Schittkowski, More Test Examples for Nonlinear Programming Codes. V, 261 pages. 1987.

Vol. 283: G. Gabisch, H.-W. Lorenz, Business Cycle Theory. VII, 229 pages. 1987.

Vol. 284: H. Lütkepohl, Forecasting Aggregated Vector ARMA Processes. X, 323 pages. 1987.

Vol. 285: Toward Interactive and Intelligent Decision Support Systems. Volume 1. Proceedings, 1986. Edited by Y. Sawaragi. K. Inoue and H. Nakayama. XII, 445 pages. 1987.

Vol. 286: Toward Interactive and Intelligent Decision Support Systems. Volume 2. Proceedings, 1986. Edited by Y. Sawaragi, K. Inoue and H. Nakayama. XII, 450 pages. 1987.

Vol. 287: Dynamical Systems. Proceedings, 1985. Edited by A.B. Kurzhanski and K. Sigmund. VI, 215 pages. 1987.

Vol. 288: G.D. Rudebusch, The Estimation of Macroeconomic Disequilibrium Models with Regime Classification Information. VII, 128 pages. 1987.

Vol. 289: B.R. Meijboom, Planning in Decentralized Firms. X, 168 pages. 1987.

Vol. 290: D.A. Carlson, A. Haurie, Infinite Horizon Optimal Control. XI, 254 pages. 1987.

Vol. 291: N. Takahashi, Design of Adaptive Organizations. VI, 140 pages. 1987.

Vol. 292: I. Tchijov, L. Tomaszewicz (Eds.), Input-Output Modeling. Proceedings, 1985. VI, 195 pages. 1987.

Vol. 293: D. Batten, J. Casti, B. Johansson (Eds.), Economic Evolu tion and Structural Adjustment. Proceedings, 1985. VI, 382 pages 1987.

Vol. 294: J. Jahn, W. Krabs (Eds.), Recent Advances and Historica Development of Vector Optimization. VII, 405 pages. 1987.

Vol. 295: H. Meister, The Purification Problem for Constrainec Games with Incomplete Information. X, 127 pages. 1987.

Vol. 296: A. Börsch-Supan, Econometric Analysis of Discrete Choice. VIII, 211 pages. 1987.

Vol. 297: V. Fedorov, H. Läuter (Eds.), Model-Oriented Data Analy sis. Proceedings, 1987. VI, 239 pages. 1988.

Vol. 298: S.H. Chew, Q. Zheng, Integral Global Optimization. VII 179 pages. 1988.

Vol. 299: K. Marti, Descent Directions and Efficient Solutions ir Discretely Distributed Stochastic Programs. XIV, 178 pages. 1988.